FINAL
FRONTIER

ALSO BY BRIAN CLEGG

*Extra Sensory: The Science and Pseudoscience of Telepathy
    and Other Powers of the Mind*

Gravity: How the Weakest Force in the Universe Shaped Our Lives

How to Build a Time Machine: The Real Science of Time Travel

Armageddon Science: The Science of Mass Destruction

Before the Big Bang: The Prehistory of the Universe

*The God Effect: Quantum Entanglement, Science's
    Strangest Phenomenon*

*Light Years: An Exploration of Mankind's Enduring Fascination
    with Light*

FINAL FRONTIER

THE PIONEERING SCIENCE AND
TECHNOLOGY OF EXPLORING
THE UNIVERSE

BRIAN CLEGG

St. Martin's Press ♒ *New York*

www.stmartins.com

Library of Congress Cataloging-in-Publication Data

Clegg, Brian.
    Final frontier : the pioneering science and technology of exploring the universe / Brian Clegg.
        p. cm.
    Includes bibliographical references and index.
    ISBN 978-1-250-03943-9 (hardcover)
    ISBN 978-1-250-03944-6 (e-book)
    1. Astronautics—Popular works.   2. Interplanetary voyages—Popular works   3. Outer space—Exploration—Popular works.   I. Title.
    TL793.C629 2014
    629.4—dc23

                                                    2014010056

St. Martin's Press books may be purchased for educational, business, or promotional use. For information on bulk purchases, please contact Macmillan Corporate and Premium Sales Department at 1-800-221-7945, extension 5442, or write specialmarkets@macmillan.com.

First Edition: August 2014

10   9   8   7   6   5   4   3   2   1

FOR GILLIAN, CHELSEA, AND REBECCA

CONTENTS

ACKNOWLEDGMENTS

With thanks as always to my editor, Michael Homler, for his help and support, and to all those who have provided me with information and assistance—you know who you are. One particular name needs mentioning, though—the late, great, wonderfully eccentric Patrick Moore, who helped so many, including me, develop an interest in astronomy and space travel.

FINAL
FRONTIER

I.

NEW PIONEERS

IIIIIIIIIIIIIIIIIIIIIIIIIIIIIIIIIIIIIIIIIII

The Earth is the cradle of humanity, but mankind cannot stay in the cradle forever.

—Russian rocket scientist Konstantin Tsiolkovsky
in a letter to an unidentified recipient (1911)

As most of us get on with our day-to-day lives, the concept of being a pioneer seems entirely alien. The closest we get to pioneering may be the choice of a different recipe for dinner, trying out that edgy new show, or being the first person from our town to visit an obscure foreign country. Yet the urge to boldly go where no one has gone before is a fundamental part of the human spirit—however much it has become a cliché, and however watered down it may be in reality. There will always be those who are prepared to risk everything to open a new frontier, whether they will be regarded as heroes or candidates for a mental health assessment.

In history, many of those people who have ventured out across the world were looking for riches or territory to claim. Others simply wanted to be the first to a new location—or to expand the sum of human knowledge. To boldly *know* what no one has known before. Yet they have all shared an inability to stick with the status quo. They all had a very human urge to push the boundaries. And what greater frontier to open up than the very limits of the Earth, taking us into space?

GO WEST, YOUNG MAN!

Of all the people on the Earth, it is arguable that none has had as much of that frontier spirit, that urge to "Go West, young man!" and be a true pioneer, than the settlers of North America. This was encouraged by a unique opportunity. The first Europeans to reach America faced a vast and largely unexplored land that lacked the out-and-out hostility of the Australian outback, but instead always offered the promise of a new, potentially wealth-creating vista ahead. In part this spirit also seemed to reflect the kind of individuals who were prepared to take on the dangerous leap into the unknown that was involved in crossing the Atlantic and entering a new world. You didn't make such a huge break with everything you knew and understood if you were an unadventurous stay-at-home person.

For some, the frontier was the only hope of an independent future where they would have more freedom to practice their particular beliefs. For others it was, rightly or wrongly, a huge opportunity and an adventure. (I am well aware that these new "unexplored" lands were often already, if sparsely, occupied. But for the purposes of identifying the pioneering drive, this was not then an issue.) Although I am sure there is no such thing as a "frontiersman gene," there seems to be some kind of inbuilt tendency to want to be a pioneer that was part of the forging of the American spirit. But what has happened to that tendency in a modern, interconnected world?

It would be sheer ignorance to say that there is no opportunity for exploration left anywhere on our planet today. There are territories where human beings have only scratched the surface. There are vast sections of the oceans in particular that have still to be explored, and still some regions of the Earth that are relatively lightly known. What's more, there is always the second level of exploration where we go from simple discovery to a more detailed understanding. As Nobel Prize–winning physicist Steven Weinberg has pointed out, the loss of "terra incognita" from medieval maps does not mean that there is nothing more to discover: "In the middle ages Europeans drew maps of the world in which there were all kind of exciting things like dragons in unknown territories. Nobody knew what was at the Antipodes." Yet the world without "here be dragons," Weinberg argues, is not a boring place. He believes that it is better to know those fundamentals, and still to have lots of interesting detail to explore. "If we had the fundamental laws of nature tomorrow, we still wouldn't understand consciousness. We wouldn't even understand turbulence . . . That's an outstanding problem that has been with us for almost two centuries and we're not very close to a solution."

Weinberg is drawing a parallel between exploring the world and knowing more about physics, but his point that there is plenty left to discover after the basic pioneering has finished continues to apply to our actual world, bereft though it may be of dragons. Even so, having a true frontier to open up, one that is genuinely new and more alien than anything we have ever explored before, will always be an attraction. The sea may be huge, but it has not got the possibility of setting up permanent colonies that goes hand in hand with the frontier. And as the Earth's unknown territories have one by one been crossed off the checklist, it is natural enough to turn our eyes upward, to look out to space, where there is enough room for pioneering to last the whole lifetime of a species and more.

MORE THAN SCIENCE

At the moment we tend to think of space exploration as a purely scientific endeavor—and it certainly can be that. But it is a mistake, not surprisingly one that is often made by scientists, to assume that scientific discovery is all that matters. The frontiers of space are just as susceptible to those other reasons for pioneering as opening up land-based frontiers ever were. As we will discover, manned missions into space rarely make sense from a simple cost/benefit analysis of their scientific worth. They have always been, and always will be, more about the other aspects of pioneering than anything else. Science is a secondary aspect of reaching out beyond our atmosphere.

This is certainly true of our experience to date. The first, tentative steps into space were the result of political posturing. The USSR got the first satellite into orbit and put the first human being into space. In response, the United States had to find an opportunity to produce the true pioneers, making the first Moon landing a hugely attractive proposition. Yes, there was some scientific work done in the process. We learned a lot from the Apollo landings and their precursor missions. But the whole context of the race to the Moon, and the way that lunar flights ceased after *Apollo 17* in 1972—only the sixth manned landing on our natural satellite, and remaining our last visit with a crew to this day—made it clear that it was all about getting there first, rather than any drive for scientific discovery.

Since then, the human race has hardly done a lot to push back the frontiers of space as far as human exploration goes. Robotic substitutes like the Mars rovers and the Voyager missions reaching the far extremities of the solar system as this book is written, have certainly opened up aspects of our solar neighborhood for us as never before. And telescopes, particularly space telescopes like the Hubble and Planck satellites, have done wonders for us in terms of the kind of remote exploration made possible by the steady progress of light across the universe. But we haven't been out there, pioneering as human beings.

EXPENSIVE EXPANSION

The clear difference between spreading out across space and the opening up of the Western frontier in the United States is the horrendous overhead of breaking out of the prison formed by the gravity well of the Earth. At the moment, this is an incredibly expensive business. According to NASA it costs around $10,000 to get a pound of material into space. Given an average weight for an American adult of 180 pounds, that is $1.8 million per person—and that's without the far greater weight of all the support equipment and resources necessary to keep that person alive. It may have been hard and dangerous to become a prospector in the American West in the nineteenth century, but you didn't need to be a multimillionaire to do it.

One of the prime movers in getting pioneers out to explore and tame this final frontier is going to be finding ways to reduce the cost of getting a human being into and around in space. This is both about finding new technologies to get us away from the Earth but also about thinking laterally and realizing that not everything involved in a mission has to come from Earth in the first place, making use of resources that are already out there.

If we can justify the expense, the reasons for getting to the new frontier remain the same as they ever were. As is clear from the current Chinese space program, politics remains a very significant driver. It may be less so in the United States at the moment—since the fall of the USSR, there really hasn't been the concept of a space race—but as Chinese activity builds it may be that once again the U.S. government will feel the need to flex its muscles and make it clear who has the technological supremacy—and in principle also who has a military foothold in locations where gravity alone has the potential to turn a lump of rock into a more powerful weapon than a nuclear bomb.

However, if the price is right, we will see all the other traditional drivers for pioneering coming into play. Ever since the early days of science fiction there has been the concept of the space miner—the deep-space equivalent of the forty-niner, who scours space, typically cruising around the asteroid belt, prospecting for rare minerals, or even water. With cheap enough spaceflight, the only real difference between the spacer and the early prospectors is that we put a higher value on human life these days. Space mining would be extremely hazardous, so on top of the energy costs will come safety costs too. That is unless we are prepared to accept that those who volunteer for such jobs would be prepared to take on significantly higher risks than we allow on the surface of the planet. In return, such miners could expect much greater rewards for their effort than anyone undertaking a similar job back on Earth.

And then there are those who seek a better life.

A LIFEBOAT FOR THE WORLD

Our population is rising. Developing nations are rightly looking at the vastly greater consumption per head of the West and want the same for themselves. More cars. More travel. More possessions. More energy consumption, burning yet more fossil fuels. And the rest of us show no sign of slowing down our frantic consumption. However, it is clear that the Earth's resources and space are both limited. One of the biggest drivers to explore the new frontier has always been the opportunity to build your own homestead, to establish personal space for an individual or a family away from the overcrowding and oversight of the city. In the very long term, if the human race still exists for long enough into the future, we will need to get out into space to continue to thrive.

As astronaut John Grunsfeld, a veteran of five space shuttle missions, put it at the Humans 2 Mars Summit at Washington, D.C., in May 2013: "Single-planet species don't survive. That's a pretty sound theorem—just look at the dinosaurs. But we don't want to prove it." In a sense, Grunsfeld's words are a truism. Given we only know of one inhabited planet, then inevitably any species that has died out will show that "single-planet species don't survive." To be fair to the dinosaurs, they managed to stay around for over 160 million years more than we have so far. In sheer survivability terms, we have a long way to go to catch them up. But any planet has a limited lifetime, and should the human race survive long enough, there will come a point when we need to get beyond being a single-planet species if we are to continue our existence. (I ought to stress "should the human race survive long enough." A more accurate truism is that species don't last forever. It is hard to imagine the human race would still exist unchanged after millions of years.)

In about 5 billion years the Sun will expand to become a red giant, flowing out into the solar system as it fluffs up to occupy a volume that is bigger than the Earth's current orbit. The Earth could in principle survive this, as it will have drifted out from the Sun by then and won't be eaten up, but it will not have been inhabitable for billions of years. This is because the Sun is gradually becoming hotter. We only have another billion years or so to go before it is beyond our ability to survive on the Earth. If human beings, or more likely our distant descendants, still exist in a billion year's time, we will need to have left the Earth to live on the final frontier whether we like it or not.

Physicist Stephen Hawking, never one to avoid a dramatic headline, has suggested that we have a shorter-term imperative to get out into space if we want to continue surviving as a species. "We are entering an increasingly dangerous period of our history," Hawking commented in an interview. "Our population and our use of the finite resources of planet Earth are growing exponentially, along with our technical ability to change the environment for good or ill. But our genetic code still carries the selfish and aggressive instincts that were of survival advantage in the past. It will be difficult enough to avoid disaster in the next hundred years, let alone the next thousand or million." The scientist believes we will soon render the Earth uninhabitable, making manned space exploration essential.

THE ULTIMATE FRONTIER

It isn't necessary to go to such drastic extremes and conjure up dystopian scenarios, though. That same drive that pushed settlers across America is likely to see colonists taking on the solar system, and perhaps beyond, on a shorter timescale than our ability to render the Earth uninhabitable. Whether we are talking of free-floating, constructed living environments in space or taking on the Moon or Mars, the pressures of life on Earth are likely to make it more and more likely that some will move away from our home planet within centuries or even decades. These first brave explorers (some might say foolhardy) may not survive. Many of the early Europeans who left for new territories to the East and West didn't. But they will have opened the way for larger-scale colonization to come.

In the short period of time that humans have been taking a scientific view, our picture of the universe has expanded immensely. Until a few hundred years ago, the usual idea of the universe was simply the solar system. The scale they imagined stopped around the orbit of Saturn—less than a thousandth of a light-year across. As our astronomical equipment has improved, so has our awareness of the scale of habitat that surrounds us. Until the early twentieth century the Milky Way, around 100,000 light-years in diameter, was considered to be the whole universe. We

now know that our nearest neighbor among large galaxies, the Andromeda Galaxy, is 2.5 million light-years away, and the universe itself is at least 90 billion light-years across, featuring an uncountable number of planets and other smaller bodies.

Even with the limited understanding of the ancient Greeks, though, the possibilities of greater stretches of frontier, waiting to be conquered, proved an attraction. According to the first-century Greek historian Plutarch, Alexander the Great was told by his friend Anaxarchus that there were an infinite number of worlds out there, causing Alexander to begin crying. When asked why he was crying, Alexander replied: "Do you not think it a matter worthy of lamentation that when there is such a vast multitude of them, we have not yet conquered one?" The idea of space as the ultimate goal for exploration—and, yes, perhaps conquest—is nothing new.

The kind of exploration of the final frontier that we can look forward to has yet to happen in reality. The cost of getting up into space has held us back. But the difficulties involved have not been able to suppress the exuberant human imagination. In novels, short stories, movies, and TV shows, the whole business of space pioneering has been examined and critiqued for decades. Science fiction is not and never has been a good predictor of the future, but it is both the nursery that has nurtured the ideas of space travel and the inspiration for some of the best real ideas in the field. So where better to start than on the frontier of this speculative art form?

2.

SPACE OPERA

||||||||||||||||||||||||||||||||||||||

It is natural that SF should be symbolized by the theme of space-flight, in that it is primarily concerned with transcending imaginative boundaries, with breaking free of the gravitational force which holds consciousness to a traditional core of belief and expectancy.
—*The Encyclopedia of Science Fiction* (1999)
John Clute and Peter Nicholls

From the first attempts to describe getting out into space that could truly be described as science fiction, it was obvious that gravity would pose a significant challenge. This barrier to reaching the frontier was approached in very different ways by the two nineteenth-century masters of the genre, Jules Verne and H. G. Wells. Each would write novels in which they would send human beings to the Moon—but there the similarity of their approach ended.

VOYAGES OF THE MIND

Verne and Wells weren't the first, of course, to write about venturing into space, but their predecessors were happy to deal more in fantasy than in science. When, for example, Cyrano de Bergerac wanted to send his eponymous character off into space in his novel *L'Autre Monde: ou les États et Empires de la Lune* (*The Other*

World: or the States and Empires of the Moon) he proposed making use of the observation that the Sun made dew disappear as a means of motive power. There was something about the Sun's rays that apparently drew the fluid off the Earth toward that heavenly body. So, Cyrano, proposed, why not capture dew in bottles, attach the bottles to your astronaut with strings, and wait for the Sun to draw the dew into space, taking the attached traveler with it? It was a plan, if not exactly scientific.

Admittedly Cyrano went on to describe something a little closer to a rocket, where a group of soldiers, unimpressed by his attempts at harnessing Sun power, attach a collection of fireworks to his space machine and send him flying once more. But there was no real attempt here at thinking through a scientific approach. The same could be said for the earlier work of English historian and bishop Francis Godwin, who wrote the story *The Man in the Moone* in the 1620s, in which he described attaching a series of lines to a flock of imaginary birds called gansas. These mythical beasts, we are told, migrate to the Moon every year. The one distinct piece of value of Godwin's work is that he emphasized the then excitingly modern idea that gravity was a force of attraction rather than a natural tendency to head for the center of the universe—so he mentioned that he experienced a loss of weight while in space.

An even more bizarre method of getting to the Moon was suggested by the astronomer Johannes Kepler, whose work on planetary motion became an essential foundation for understanding navigation of the solar system. At around the same time as Godwin was writing, Kepler produced a piece of fiction called *Somnium* (the name, meaning "dream" makes it clear that this was a fantasy). Kepler has his hero transported across a bridge of darkness by demons who live on the Moon and cross over this mystical span during a solar eclipse. However, these seventeenth-century fantasies were enough to set the seed that would come to fruition when true science made its way into the world of fiction in the nineteenth century.

BLASTED INTO THE HEAVENS

As it happens, neither Verne nor Wells came up with a very practical method of making the journey to the Moon in their novels—but then, it is interesting that both authors, usually serious in tone, took a humorous approach to the subject. Verne came up with a more realistic view that it would be necessary to propel a capsule out of the Earth's gravity well using force, but his *De la Terre à la Lune* (*From the Earth to the Moon*) was based on a flawed concept. The story features the Baltimore Gun Club and its members' plan to produce a huge cannon, so powerful that it could shoot a shell containing human cargo all the way to the Moon. It's a sobering reflection of Verne's idea of American culture that this project was not devised to explore new worlds but to act as a distraction, because the Gun Club members were bored because there was insufficient warfare being waged to keep them amused.

Verne was detailed in his description of the technology. The projectile was made of the then decidedly up-to-the-minute material aluminum, 2.7 meters (108 inches) in diameter and weighing 8,730 kilograms (19,250 pounds). The gun, known as the Columbiad, was 274 meters (900 feet) long and made of cast iron, running straight down into the Earth. (In fact it would have been better to shoot at an angle to make use of the Earth's spin to lower the escape velocity.) And the propulsion was achieved with 181,000 kilograms (400,000 pounds) of gun cotton. With ironic coincidence, the Columbiad was constructed in Florida, quite near Cape Canaveral, after an argument is played out between Texas and Florida over where to site the space center, reminiscent of the political rather than logical decision to place NASA's control and training centers so far from Florida in Houston, Texas.

In reality, the device that Verne imagined was an entirely impractical way to put men on the Moon. Unlike a rocket, which is under power during flight, the cannon would have to get its projectile all the way up to escape velocity, the speed needed to escape Earth's gravity—around 11.2 kilometers per second (7 miles per second)—before the capsule left the barrel of the gun and was in free flight. With any practical length of barrel this would involve a huge acceleration. Even if the gun itself were capable of it, any humans on board would be turned into mush. The acceleration involved would be impossible to survive. With, for instance, a much longer 10-kilometer (6.2-mile) gun barrel, the acceleration would be the equivalent of over 600 g—600 times the pull of gravity on the Earth's surface—many times the survivable level.

As we will see later on, variants on the space cannon are not impossible if you are faced with less gravity, and hence less acceleration required to escape from the gravity well. But for the Earth, this approach was never going to be realistic, something Verne no doubt realized. He does make a passing reference to "countering the effect of the shock of the departure," but his mechanism for doing so is little more than a deus ex machina. His hero Barbicane simply says that water will "act as a spring," presumably cushioning the impact—an entirely inadequate precaution for the tremendous forces the travelers would experience. Later, in the sequel *Autour de la Lune* (*Around the Moon*), Verne contradicts himself, saying that the capsule was lined with "strong springs and partitions to deaden the shock of departure," which really is no better solution for survivability.

This didn't stop the book from being a great success and having a number of spin-offs, notably and somewhat bizarrely the world's first science fiction opera *Le Voyage dans la Lune* by Jacques Offenbach, better known for his popularization of the cancan. More significantly, it also inspired the first epic science fiction movie, *A Trip to the Moon*, which cinematographer Georges Méliès made more of a comic turn than the original, showing the projectile landing in the Man in the Moon's eye. Even so, we shouldn't underestimate the impact of this film, because it was, by standards of the day, a major undertaking, running for twenty-one minutes at a time when very few movies were more than two minutes long.

THE ANTIGRAVITY SOLUTION

By comparison with the brute force of Columbiad, H. G. Wells's mechanism leaves his astronauts suffering no unpleasant feelings at all—because they simply float away from the Earth's surface in a machine that requires no exertion of energy to get them out of the Earth's gravity well. If this sounds too good to be true, it is. In *The First Men in the Moon*, Wells makes use of a ship covered in blinds that can be opened and closed. Each blind has a coating of cavorite, a substance devised by the antihero Mr. Cavor. This is an imaginary artificial substance that is impervious to the pull of gravity. As Wells puts it:

> Now all known substances are "transparent" to gravitation. You can use screens of various sorts to cut off the light or heat, or electrical influence of the sun, or the warmth of the earth from anything; you can screen things by sheets of metal from Marconi's [radio] rays, but nothing will cut off the gravitational attraction of the sun or the gravitational attraction of the earth. Yet why there should be nothing is hard to say.

This is fine as a means of propulsion for a work of fantasy, but frankly is of little use in giving us a hint as to how the task might be achieved in the real world. There is no known substance that is not transparent to gravity, and there is never likely to be. Nothing will shield against gravity. If such a material existed it would be just as valuable on Earth as it would be to propel a space vessel. It would make it possible to build a perpetual motion machine, producing energy from nowhere.

All you would need is to have a wheel, a little like a water wheel, with the same side of each paddle coated in your magic gravity barrier. As that side of the paddle would not be attracted by the Earth, the paddles with the other side facing downward would be pulled downward, turning the wheel with no source of energy. Cavorite offers us a wonderful vision of limitless free energy—and all we know of physics tells us that this could never be possible.

SPACE ON THE SILVER SCREEN

We'll come back to books, but the making of Méliès's moving picture trip to the Moon is the ideal point in history to take an excursion into moving pictures. Strangely for such a technological medium, the movies have always lagged far behind the written word in their portrayal of science fiction. Science was slower to work its way into attempts to portray space travel on celluloid, where for a number of years films about space would be pure fantasy that had more Cyrano de Bergerac about them than any consideration of practical methods to escape the Earth.

Often the mechanism for getting into space was nothing more than a dream, while in Gaston Velle's 1906 short *Voyage Autour d'une Étoile* (*Voyage Around a Star*), the protagonist floats off into space in a giant soap bubble, somehow managing to cross interstellar distances and return in the six-minute duration of the film. If all else fails in the early movies, then it's back to Wells and some mysterious chemical concoction is introduced that gives the power of antigravity to get the action started.

The missing link between the early fantasies and the more realistic later attempts is probably the 1917 Danish film *Himmelskibet* (technically *Skyship*, but given a range of English titles from *The Airship* to *400 Million Miles from the Earth* and *A Trip to Mars*). Here it is no longer enough to float magically through space (or to make use of the occupant-splattering effect of a space cannon). In *Himmelskibet* we meet a real-life spaceship. Of sorts. The ship looks like a cross between an airship and a biplane, and is propelled through space by the usefully vague concept of "a force that we discovered." In essence we are still dealing with magic, but at least it is magic that is given a gloss of science and technology.

To have a movie that even vaguely approximated to reality, the viewing public had to wait for Fritz Lang's *Die Frau im Mond* (*The Woman in the Moon*, but sometimes titled *By Rocket to the Moon*). Lang had already demonstrated his science fiction credentials with the astonishing three-hour 1927 epic *Metropolis*, featuring a humanoid robot and vast armies of workers employed in the ultimate nightmare of an over-industrialized city, but the 1929 *Die Frau im Mond* saw the first venture on celluloid of rockets being used to get human beings into space.

Unlike the remarkable *Metropolis*, *Die Frau* was no great shakes as a movie, but Lang brought in scientific advisors including rocket scientist Hermann Oberth, who would later work on the V-2 rocket-powered weapon, and whose students included Wernher von Braun. Oberth knew his stuff, so much so that the film was later withdrawn by the Nazis and the model rocket used in the movie destroyed because it was all far too close to the work then being undertaken on the V-2 (see page 52). It was in *Die Frau* that a scientific approach to space travel finally began to replace pure fantasy. Not only was the motive power of the spaceship a rocket, but this was a rocket using multiple stages to avoid having to carry too much mass into later parts of the journey, an essential part of many real-life space missions.

In an example of life imitating art, one other small feature of this movie would become a reality when NASA began its real-life equivalent. Lang wanted to build tension leading up to the launch of the rocket in his movie and featured a countdown of seconds before the takeoff. The countdown rapidly became a part of the fictional culture of space travel, so much so that real NASA launches would also make use of it. This speaks volumes for the degree to which space ventures have always been as much about the audience on Earth and about politics as they are about what happens to the astronauts.

The decade that followed the premier of *Die Frau im Mond* brought little progress in movie space travel. One of the better years was 1936, which saw the ludicrous puttering spaceship of Flash Gordon, and the attempt to film H. G. Wells's sprawling and incredibly tedious "future history" book *The Shape of Things to Come*. Although an epic for its day, and in some ways impressively aware of contemporary events—*Things to Come* portrays the Second World War beginning in 1940—the final scene of the movie takes a huge retrograde step in featuring a departure into space using of all things a Jules Verne–style space cannon.

MAKING THE MOVIES REAL

It wasn't until 1950 that movie space travel was set back on the course started by Fritz Lang and Hermann Oberth with George Pal's lavish production *Destination Moon*. This brought a documentary-style realism to the science fiction adventure. Even the setting was right for the time. The trip to the Moon was set in the context of a Cold War race to get there first, because of the Moon's strategic military importance. Once again Hermann Oberth was on hand to advise.

The ship itself was a classic rocket design from pulp magazine covers, like a more shapely, center-bulging version of a V-2, and able to land on the Moon precariously upright—but if anything Pal put too much effort into trying to give a realistic feel to his fictional mission. The result was a movie that is decidedly dull to watch.

The specific technology in *Destination Moon* is less realistic than the feel of the movie. The rocket is "atomic powered" (almost an essential in the post-Manhattan Project fervor for all things atomic), though it is not really clear what this means, and is a single stage rocket with no jettisoned stages, something that has never been achieved in getting a real rocket of any size into deep space. The venture in the film is a private one, reflecting more the twenty-first century's possibilities for commercial space exploration than the NASA-dominated landscape of the real 1960s race to the Moon.

Pal also showed us a spaceship leaving Earth to act as a survival vehicle in his 1951 movie *When Worlds Collide*. In this melodrama, the Earth is about to be destroyed in a collision with a rogue star called Bellus. The spaceship does not have too far to go as Bellus handily brings with it a planet called Zyra, which passes close to the Earth. The ark-like escape vessel, once more privately funded, is similar to the *Destination Moon* ship in design, but differs in being launched along a track that starts with a long horizontal run and curves up at the end, not unlike part of a modern roller-coaster ride—totally impractical in reality. Such a structure may be feasible in a low-gravity environment like the Moon, where it is more likely to feature as a mass driver (see page 59), but it would be impractical to build a structure like this on Earth that could withstand the forces involved in accelerating a massive rocket to escape the Earth's gravity well.

In parallel with the more realistic rocket approach, other movies of the period had a fascination with flying saucers, arguably more related to the early "magic" portrayal of space travel, as these rarely if ever had anything more than a handwaving mechanism for flight and did not parallel any practical development of our ability to take on the final frontier. Though there would be plenty such alien vehicles—or in the hands of future humans in *Forbidden Planet* (loosely based on Shakespeare's *Tempest*), and even arguably in *Star Trek* (though admittedly the warp drive doesn't seem quite as based on magic now—see page 250), *Star Wars*, and all their imitators—there have been relatively few movies that have provided anything close to a realistic vision of a future in space for the next hundred years or so.

MOVIES THAT SHOW THE WAY

There are really only three movies and one TV series worth mentioning. Without doubt, the most remarkable of these was *2001: A Space Odyssey*. Released in 1968, a year before the first Moon landing, it is easy now to pick holes in *2001*. With hindsight, the timescale for the achievements shown in the movie was ludicrously optimistic. To think that we would have a huge wheel-shaped space station, a Moon base, and the ability to send a manned expedition to the moons of Jupiter by the first year of the twenty-first century was wildly unrealistic. And the film is occasionally undermined by the details that are included to show how grounded it is in reality. In *2001* we see Pan Am space shuttles carrying passengers to the great rotating wheel of a space station—ten years after Pan Am collapsed. Not to mention impressive video telephones operating back to Earth—but no cell phones.

However accurate these quibbles, to pick at *2001* in this way is to miss how much of a step forward it was in the portrayal of the conquest of space. Gone were the missile-like spaceships of the V-2 heritage to be replaced by more practical shuttles, while the immense

Jupiter mission ship, *Discovery*, has no pretense of being aerodynamic because it is never intended to enter an atmosphere. Instead it is a long, messy conglomeration of units, clearly assembled in space as it could not withstand full Earth gravity, let alone the strains of heaving it up from the surface into orbit. There is also the realization in *2001* that undertaking large-scale exploration in the solar system is going to require serious infrastructure in space and quite possibly on the Moon. Compared with the space station in *2001*, the current International Space Station (ISS) is like putting a homemade shed assembled from scrap alongside a mansion.

The second significant movie is the much less well-known *Silent Running*. Released four years after *2001*, and directed by Douglas Trumbull who had been in charge of the special effects on the earlier film, *Silent Running* portrays a future when the Earth's plant life has been wiped out by humanity's rapacious misuse of the planet. A remnant of the Earth's forests have been kept alive in domes carried by space freighters, located (for no good reason other than the opportunity this provides for impressive special effects) out beyond the orbit of Saturn.

The movie itself is heavily flawed, with irritatingly twee robots and a self-sacrificing hero who eventually destroys everything on the ship except for a single dome that is left to float off into space maintained by one of the robots, carrying Earth's precious remains. However, *Silent Running* does pick up on a number of potential key points for space exploration. There is the possibility of needing to salvage an ecosystem that can no longer survive on the Earth, and the interesting idea of having a biosphere, a relatively small but balanced environment in a dome, rather than the totally artificial environment we associate with spaceships.

In *Silent Running* and *2001* everything may not be rosy, but the technology is pretty sleek and unlike any of the real space equipment we have used so far. The third significant film is the 1972 Russian movie *Solaris* based on the book by Stanislaw Lem, remade with George Clooney in the starring role in 2002. *Solaris* is important because it features more realistic, clunky technology, and examines the tensions—mostly psychological—of being trapped in a space station. The personnel experience loneliness, isolation and yet at the same time are forced into close proximity with others. These are real factors that will play strongly, particularly in long-range missions, and though *Solaris* has mystical undertones and strange replicated humans that get in the way of this theme, it still is an important milestone for the genre.

Another premise that comes through strongly in plenty of written science fiction, but rarely in the movies is why I wanted to include one TV show—*Firefly*, made by Joss Whedon, the creator of *Buffy the Vampire Slayer*. *Firefly* was cancelled before the season was completed (though Whedon closed off some of the threads left open in the story in the sequel movie *Serenity*), which was a shame as this show provided an excellent exploration of the idea of space as frontier.

In the show, the ship *Firefly* is a freighter, hauling materials that may or may not be legal from place to place on the frontier of human occupation in space. This is a scenario that is inevitably strongly influenced by the American West of the nineteenth century. We see an environment where the gun is often a necessary accompaniment to a deal and where intersystem traders have to be prepared to think on their feet and respond to the unexpected. What *Firefly* does well is portray the rough and ready nature of pioneering life. This is no slick *2001* world. The machines that keep the traders alive are forever being patched up and kept running on a shoestring. The characters constantly encounter the harsh economics of life and death that accompany existence on the frontier.

A LITERARY FUTURE

By comparison with movies and TV, the book world presents us with a much richer picture of future exploration. Much of this will come up later, but science fiction novels have the opportunity to uncover the political underbelly of space exploration, and to give that frontier experience a far deeper resonance. Written science fiction typically has more realistic characterization and presents far more of the potential problems and opportunities of life in space. Before we take a leap into the real world, let's briefly look at a handful of examples.

Technically the first example of a thoughtful analysis of space travel wasn't science fiction so much as futurology, though the term hadn't been invented in 1929 when *The World, the Flesh and the Devil* was written by the pioneer of X-ray crystallography, John Desmond (JD) Bernal. You might think from the subtitle "An Enquiry into the Future of the Three Enemies of the Rational Soul" that this was a theological or philosophical tract, but Bernal examines in some detail the potential for human space travel in the future, coming up with some surprising insights. Generally speaking "futurology" is nothing more than science fiction without a decent storyline, but at times Bernal's writing verges on the prophetic, especially when you consider that this book was written when most of the pioneers of the rocket age were children. Wernher von Braun, for instance, was fifteen at the time of publication.

Bernal believed that limitations of living space on Earth and, in the long term, the geological changes on the planet (presumably he had in mind future ice ages) would lead us to look to a new frontier. "Already ambition is stirring in men to conquer space as they conquered the air," he comments. Remember this is 1929, thirty years before the first unmanned venture into space. Bernal thought that the rocket would continue to be the primary means of power (itself a huge novelty back then), though he moved away from crude chemical energy. By the time rockets became practical, Bernal thought, they would be boosted into space by some form of beamed electromagnetic energy. He realized that one of the most difficult aspects of space travel would be landing on another planet or returning to Earth, so he suggested that initial exploration should be simple flybys. And he encouraged the development of solar sails, a means of propulsion that would not be made practical until the twenty-first century.

Although Bernal wasn't aware of the difficulties that arise from long-term weightlessness, and so did not attempt to provide his hypothetical ship with artificial gravity, he did come up with the concept of a ship large enough to live on for many years. This was a 10-mile diameter, mostly hollow shell with 20,000 to 30,000 inhabitants who lived by converting any incoming light to energy using either solar cells or a modified version of chlorophyll. (Again, Bernal is well ahead of his time in considering the possibility of artificial photosynthesis.) And, eventually, he believed that we would see the conversion of human beings into cyborgs who could better survive the rigors of life in space. Star Trek fans may well think Bernal had prefigured the Borg, but here was someone setting out to imagine a distant future, identifying most of the problems that would face real space exploration with remarkable accuracy.

HISTORY AS FUTURE

As we explore science fiction novels more generally, the nearest to any possible reality tends to occur in stories featuring asteroid miners. Once thought to be an exploded planet, we now know that the asteroids between Mars and Jupiter are simply debris left over from the coalescence of the solar system (there is far too little mass in all the asteroids put together to make up a full-sized planet). It was realized fairly early on that this floating field of debris had the potential to be a rich source of minerals and rare metals. In fact one of the earliest books that could fairly be described as science fiction is the 1898 *Edison's Conquest of Mars*, a collection of stories originally written as a serial by Garrett P. Serviss in which an attack on Mars by Earth spaceships discovers an asteroid being mined for gold by the Martians.

A fair number of later novels used asteroid mining as a background, often drawing on the model of the Californian gold rush but writ large in the sky. Asteroid miners are generally rough and ready, prepared to cut corners, but contrast favorably in their youthful vigor with the jaded, over-sophisticated remnants of Earth civilization. Although not a miner himself, Han Solo in the Star Wars movies typifies the kind of character that was often found out and about in the asteroid belt.

Probably the best example in written science fiction comes in the Belter part of the large Known Space series by veteran science fiction writer Larry Niven, where echoes of the cold war are portrayed in the tension between the superpowers of Earth and the [asteroid] Belt. Although there is always the possibility of conflict, the two are interdependent—Earth leaning on the Belt for the minerals it needs to replace the long-exhausted terrestrial mines, and the Belt relying on the sophistication of Earth's processed goods and food. There is something of a reflection also of the contrast between the young American nation (the Belt) and the British nation (Earth) that spawned it, but is now surpassed by it for energy and enthusiasm. Though there are echoes of this type of story in *Firefly*, it's a motif that has rarely made it to the big screen.

ENCOMPASSING THE STARS

A second rich vein comes with the next level of exploration, the ultimate frontier—reaching for the stars. While science fiction is perfectly capable of waving a magic wand and overcoming physical barriers like the light speed limit, a fair number of novels have been set in a universe where there was no breaking of Albert Einstein's restriction—where nothing can move faster than light. With the nearest star over 4 light-years away, it would take around 114,000 years to reach it traveling at the speed of the fastest human beings yet, the crew of *Apollo 10*, who flew at 39,897 kilometers (24,790 miles) per hour relative to the Earth. Even though any practical starship would have to travel much faster than this, it may have to venture tens of times further to reach a worthwhile destination—and it would be journeying for a timescale that would only be practical in an environment where individuals could live and die on board: a so-called generation ship.

There are many examples of these in literature. Sometimes the occupants have been on the ship so long that they forget what it's all about. This can result in a collapse into barbarism, as in the early Brian Aldiss novel *Non-Stop*, where the occupants don't even know that they are on a ship. In fact surprisingly few generation ships in literature ever seem to get to their intended destination, perhaps because they were not really created by the author to do so, but to provide a good way to study humans operating in a confined space. As such, from the point of view of the writers, the journey becomes the essential part of the story, rather than reaching the destination.

One uniquely forward-thinking example of a kind of generation ship is *Cities in Flight*: an omnibus volume of a four-book series by the mostly forgotten author James Blish. It's highly unfair that Blish isn't more read today because he had some brilliant ideas. *Cities in Flight* is a strange collection consisting of a prequel, *They Shall Have Stars*, which sets up the political conditions and scientific background to the story; a young adult novel, *A Life for the Stars*, which introduces the main concepts of the storyline (the best of the books in terms of storytelling); and two far-reaching adult novels, *Earthman, Come Home* and *A Clash of Cymbals* (also known as *The Triumph of Time*), which features the end of the universe.

It is particularly striking going back to read *They Shall Have Stars* now as it is set in our present, but written in 1956. Inevitably some things date it. The way women are treated, for example, seems positively medieval. And there are the inevitable technical hiccups like imagining that Blish's twenty-first century characters would still have slide rules or copies of documents produced on duplicators with their distinctively pungent ink smell. However, the book also has some remarkably prescient aspects, bearing in mind it was written before the first satellite was launched and before the first human had ever ventured into space.

In the book we see that by the twenty-first century space exploration has stalled, with no real technical developments being made in decades—a very straightforward parallel with our current position, although in Blish's 2013, Earth has achieved manned expeditions to the moons of Jupiter, where a major science experiment is underway in the gravitational field of the giant planet. Even so, in the book there have been no major space developments since the 1980s, and there is a lot of questioning as to whether big-budget science projects are worthwhile. It's a scenario that would be familiar to any modern space scientist.

Blish addresses the need to head for the stars with stunningly imaginative science fiction. He sees two essentials to making such long journeys (even though his technology does enable faster-than-light travel)—getting communities into space, and having individuals who can live indefinitely. So the twin scientific themes of the first book are the development of antigravity technology called spindizzies that enable whole cities to lift into space, and the production of "anti-agathic" drugs that can extend life indefinitely, necessary for the immense journey times in interstellar space.

These drugs enable Blish to examine the social consequences

of having a substance that extends life, especially if there are insufficient stocks for everyone to receive it. And it features the incredibly bold concept of not just producing generation ships but whole cities full of long-lived occupants, which is truly astonishing. While this is unlikely ever to be a practical solution—although Blish uses a lot of nearly real science to explain his technology, antigravity does not seem likely ever to be available—it is certainly a very effective vehicle for examining the possibilities for an impact of large-scale interstellar departures for humanity.

THE PERILS OF TIME DILATION

The final example of useful science fiction insights into interstellar travel comes from a single book, *The Forever War* by Joe Haldeman. Intended as a specific antidote to the gung-ho militarism of Robert Heinlein's *Starship Troopers*, Haldeman's book features interstellar war like Heinlein's, but brings in the twist of special relativity. The ships in *Forever War* travel at sublight speeds, but sufficiently close to the speed of light that they experience massive time dilation. Thanks to the effects of relativity, time runs much slower on the ships than on the home world. So having been shipped off to war on a journey that subjectively may only take a few weeks or months, the troops return to find everything they knew is in the far past. No friends, no family are left alive—their only connection to reality is the other troopers. And so they sign up for tour after tour, alienated from the rest of humanity.

I don't put this in to suggest that wars in space are likely to be a major part of the future of exploration, because they aren't. Frontiers are traditionally dangerous and lawless places, and it would be surprising if justice were not fairly thin on an interplanetary frontier, but the sheer logistical task of supporting troops light-years from home suggests that if there were ever interstellar war, it would depend entirely on unmanned vessels. Starship troopers are unbelievably unlikely at best, but Haldeman's book is very useful in the way that it explores the implications of time dilation for long-distance space pioneers—and the inevitable detachment from the rest of the human race that goes with it.

Science fiction can provide us with many hints of what to look out for—but the danger of using science fiction as a guide with too much enthusiasm is to forget the importance of the "f" word. These books and movies are, first and foremost, fiction. A while ago I appeared in a short video on the science of time travel that was made to accompany the Blu-ray version of the movie *Looper*. (Probably the only time I will appear in the same production as Bruce Willis.) Although there are aspects of the science in *Looper* that are intriguingly close to reality, it doesn't get everything right: but I argued that this is beside the point. In science fiction, if there is a choice between a good storyline and good science, the science *has* to come second. Ideally you want both, but sometimes the science has to be sacrificed. It is storytelling that matters in fiction first and foremost.

If we aren't going to depend too much on fiction, we have to take our guide from real space exploration. It might seem that this has been pitifully meager. A small space station, a few manned trips to the Moon—all backyard stuff with little real expansion of the frontier. Yet that would be to miss out on the fact that we have been exploring the universe for many hundreds of years. It's just that the exploration we have undertaken has been done without ever leaving the Earth's surface. By using the medium of light.

3.

SEEING FURTHER

||||||||||||||||||||||||||||||||||||||||||||||

By the means of Telescopes, there is nothing so far distant but may be represented to our view . . . By this means the Heavens are open'd, and a vast number of new Stars, and new Motions, and new Productions appear in them, to which all the ancient Astronomers were utterly Strangers.

—*Micrographia, or Some Physiological Descriptions of Minute Bodies Made by Magnifying Glasses with Observations and Inquiries Thereupon* (1665)
Robert Hooke

The obvious place to start the history of the exploration of the universe might appear to be October 4, 1957, the date that *Sputnik 1* was launched into orbit. Yet though Sputnik was a precursor to manned expeditions into space, its true descendants were unmanned satellites, the likes of the Hubble Space Telescope and the Planck space observatory, making use of light in a tradition of visual exploration that goes way back, past Galileo's famous telescope to the very first individuals who peered up at the skies and wondered. It is light that has given us our only way to explore the universe for most of our existence.

SEEING THE FIRST LIGHT

Light finds it easy to go where we as solid, slow material beings can't. A good 99.9999 percent of all the space exploration we have ever done has been from the surface of the Earth via the light that streams in from the stars. Sitting comfortably in an observatory may not be our idea of the frontier spirit, but we should celebrate just how much exploration has been and can be done using everything from radio to gamma rays before we take the plunge into exploration in the traditional sense. (Astronomers would point out, incidentally, that their job isn't always comfortable—telescopes have to be at ambient air temperature, often chilly on mountains. To be fair, though, many modern astronomers work remotely from nice heated offices.) Our ability to see far into the universe gives us a huge advantage in exploring space over any previous land-based pioneers who did not have the same ability to see ahead—it would be silly to ignore it.

As I write, the data from the Planck satellite is just giving us the best-ever view of what is sometimes described as the "first light" of the universe. If the standard big bang model of the origins of the universe is correct, for the first 380,000 years or so of the existence of the universe, no light could pass through it. This is because it consisted of a plasma of electrically charged particles, far too capable of absorbing any passing photons to allow them to pass through. But as the relatively young universe continued to expand and cool, that collection of ions and electrons that constantly interfered with the passage of the electromagnetic phenomenon that is light, settled down as uncharged atoms and the universe became transparent.

The photons that were crossing the universe at that point have kept going ever since, for over 13 billion years. The continued expansion of the universe has meant they have lost energy—what was a white-hot glow has dropped down into the invisible micro-wave region. If the light had been produced at the time by matter it would have been heated to a healthy 3,000 kelvin (4,940 degrees Fahrenheit), around half the surface temperature of the Sun, where now it has dropped to the light output expected from a chilly 2.7 kelvin. (Such temperatures are measured in kelvin, units with the same size as a Celsius degree but with 0 at −273.15 degrees Celsius, absolute 0. This makes 2.7 kelvin the same as −454.81 degrees Fahrenheit.) This "cosmic microwave background"—sometimes called the echo of the big bang, though this is pretty inappropriate under the circumstances as it appeared 380,000 years too late—was first detected by a dish designed to pick up satellite communications in the 1960s.

The same cosmic signal was also present in the onscreen static picked up when old analog TVs were tuned between stations, the irritating snow and hiss that few realized was the oldest signal in existence. The radiation comes from all directions in the universe (which is one of the reasons it is thought to have this fundamental origin), and is almost uniform, with just tiny variants that are cleaned up to produce the familiar egg-shaped patterns produced by the COBE, WMAP, and now Planck satellites, showing the faintest image of what structure there may have been in the universe so far back.

The cosmic microwave background (CMB) has always been one of the big supporting elements of the standard big bang model, though even the big bang's first real competitor, the steady state theory was modified to match up to the CMB before the death of its main protagonist Fred Hoyle. Interestingly though, the early indications from the Planck results are that once again the standard model is going to need some patching up (for about the fourth time) before it matches the data. There are those who argue that any theory can be made to fit the data if you keep changing it every time new data comes in . . . but the big bang remains our best theory for the origins of the universe at the moment.

The very fact that we can be thinking about how well a theory matches the data on what the universe was like over 13 billion years ago demonstrates the power of light as a means of exploring the universe. At first sight the limitation of the speed of light may seem a disadvantage. It means that we have no idea what is going on beyond our neighborhood right now. In principle the whole of the rest of the universe could have disappeared and we wouldn't find out until the light stopped arriving. But look at it another way and there's a huge benefit to be had. Because light gives us a tunnel through time.

THROUGH THE TIME TUNNEL

Compare the work of the cosmologist with the archaeologist. Both are trying to discover what happened a long time ago by indirect means, without the ability to go there and undertake experiments. Neither can get hands on with the living original. But the archaeologist is limited to digging up remains that have been left to decay and deteriorate for many years (e.g., dinosaur remains that are millions of years old). By contrast, while the cosmologist equally can't get her hands on actual planets and stars, she can see directly into the past.

Light travels at a steady 300,000 kilometers (186,000 miles) per second—or to be precise, 299,792,458 meters per second, a velocity that will never change with more accurate measurement as the meter is defined as 1/299,792,458th of the distance light travels in a second. Because of the scale of the universe, this means that by looking out at distant objects we see directly how they were in the past. The photons that arrive at the Earth give us the view at the point in time that they set off, a long time ago. Take a look, for instance, at the constellation of Andromeda, located below the distinctive W of Cassiopeia. Follow the direction pointed to by the second V point of Cassiopeia by about the width of the W and, on a clear night with good eyesight you will see a little fuzzy patch.

This patch of light is neither a star nor a cloud in space, it is the Andromeda Galaxy, a huge spiral structure containing billions of stars, the nearest large neighbor to our own Milky Way. The Andromeda Galaxy is the most distant object directly visible to the human eye. It is around 2.5 million light-years away. Bearing in mind that a light-year is simply the distance light travels in a year, this makes 1 light-year 300,000 kilometers (or 186,000 miles) a second, multiplied by 365.25 × 24 × 3,600 seconds in a year, or around 9,467,280,000,000 kilometers (5,869,713,600,000 miles)—this means that the Andromeda Galaxy, at 2.5 million times this distance is distant indeed. Compare it, for instance with the feeble 380,000 kilometers (236,000 miles) to the Moon, the furthest a human has ever traveled. And we need to remember that in terms of the scale of the universe, the distance to Andromeda is itself trivial.

Realistically, unless we can develop some form of faster-than-light drive (see page 250), light will always be our principle mechanism for exploring the Andromeda Galaxy, or the more distant reaches of the universe. Even if we managed to get a ship up to half the speed of light, requiring unimaginably vast amounts of energy using current technology, it would take 5 million years to reach our neighboring galaxy. And if we think of our existing probes, the challenge is even more stark. *Voyager 1* is our furthest traveled piece of technology. At the time of writing it is beginning to leave the solar system, having journeyed 18 billion kilometers (11 billion miles). But at its current velocity of 17,000 meters per second (around 38,000 miles per hour) it would take nearly 74,000 years to reach the nearest star and 439,125,000,000 years to make it to Andromeda. Without warp speed, light is our only hope to find out about Andromeda and the further-out galaxies.

When we look out on the Andromeda Galaxy, we don't see it as it is now, but as it was 2.5 million years ago, long before human beings existed. When you go to the extreme limits of the universe, using our most advanced telescopes, when observing the most distant objects we are looking back over 13 billion years in time. Oddly these objects are not around 13 billion light-years away. Because the universe has been expanding at an accelerating rate, if you pick an object whose light has been traveling toward us for 13 billion years, like the gamma-ray burst GRB 090423, which is often mistakenly described as being 13 billion light-years away, its actual current distance from us is more like 30 billion light-years. But this does not alter the time that the light has been on its way.

This means that astronomers and cosmologists are truly looking out through a time tunnel. The equivalent facility for an archaeologist or paleontologist would be if they had a device through which they could peer and see Neanderthal man living in Europe, or watch as dinosaurs walk across the plains of America, living out their lives in view of the scientists. When exploring space we might be frustrated in our inability to touch, to have a hands-on experience, and be able to do direct experiments, but we do have the huge benefit of being able to see into the past.

A MODEL UNIVERSE

Initially the enforced distance from what was being observed resulted in a poor match between what we now believe to be out there and the understanding of early earthbound philosophers. One of the earliest scientific models of the universe was Aristotle's dating back to the fourth century BC, a picture that would remain in force for around two thousand years. Knowing what we do now, his universe seems quaintly improbable, but Aristotle was simply following the logic of what was understood about the nature of reality at the time.

His universe had the Earth at its center, the hub of reality. Around our planet (already known to be shaped like a ball) were a series of hollow spherical crystal shells, each carrying one of the planets, literally the "wanderers" of the sky. These were what we would class as the planets from Mercury to Saturn (excluding, of course, the Earth), but it also included the Moon and the Sun. In a final shell at the edge of the universe were the fixed stars. Aristotle himself did not come up with an idea of how big this universe was, but Archimedes, born around a hundred years later, and a much more practical thinker, managed to work out the scale of both this universe and of a second, hypothetical universe based on the then novel idea that the Sun rather than the Earth could be at the center of everything.

Archimedes brings this up in a strange little book called *The Sand Reckoner*, in which he attempts to calculate the number of grains of sand it would take to fill the universe. This isn't quite as bizarre an undertaking as it sounds. What Archimedes was really doing was demonstrating just how limited the Greek number system was, with a largest named value of a myriad (10,000), or at best a myriad myriads (100 million). He showed how it could be extended indefinitely to cope with vast numbers. But in order to work his demonstration, Archimedes had to first estimate the size of the universe.

With a bit of fancy geometry (he was an ancient Greek mathematician, after all) he was able to come up with a diameter for the universe of around 10 billion stades. This is a measure based on the distance around a stadium's track that was a common measure at the time, their equivalent to estimating in football fields. Embarrassingly it isn't entirely clear what distance Archimedes was using, as there were several different versions of the stadion—it is thought to have been 600 feet, but the foot was not standardized, varying from city to city throughout Greece. Roughly, though, a Greek stadion seems to have been around 180 meters (590 feet), making the size of Archimedes' estimated universe around 1,800 million kilometers (1,120 million miles) across.

This measurement puts the edge of his universe outside the orbit of Jupiter, but not quite including Saturn which really wasn't a bad guess given the very limited information available to him.

Archimedes' second estimate was based on a Sun-centered model, written about in a book by his contemporary Aristarchus. Unfortunately the original book is lost, so this is the only reference we have to this early move away from putting the Earth at the center of everything. The shift in Archimedes' geometry to cope with the new position was significant, increasing the size of the universe by a factor of 10,000, making it an even better fit to the current accepted size of the solar system (which is, in effect, what Archimedes' universe was), incorporating all the planets and the rocky Kuiper belt beyond them.

Although Aristotle's model was pleasingly simple, and there's a lot to be said for that, it had a serious problem when it came to using it to explore the workings of the universe. It didn't fit the observations made of the real universe. This doesn't seem to have worried Aristotle too much, as he always gave a higher weighting to philosophical argument than to observation (he famously declared that women had fewer teeth than men without ever bothering to count them), but for those who had a more scientific view and were attempting to understand exactly what was out there, it was clearly a problem.

TAMING THE PLANETS

The difficulty with Aristotle's universe was that some of the planets were in reality badly behaved, moving in a way that simply wouldn't work if they were fixed by their position in crystal spheres to move in a circular orbit. Where the Sun and the Moon, for instance, made steady progress through the sky, some of the planets, like Mars, seemed to suddenly change their mind and reverse direction against the backdrop of the stars, an action known as retrograde motion.

From a modern viewpoint this is easily explained. Mars and the Earth travel around the Sun at different speeds, traveling around orbits that aren't neat concentric circles. The result is that when seen from the surface of the Earth, combining the two motions, Mars seems to loop back on itself as the faster moving Earth overtakes it—but this simple explanation (which would later be one of the main arguments for the modern Copernican model of the solar system) was incompatible with the steady grinding of Aristotle's Earth-centered crystal spheres.

Although Greek philosophers were prepared to give up on the immutability of those crystalline spheres, they were not prepared to move the Earth from its special place at the hub of the universe. The whole philosophy of matter was based on the idea that heavy things had a natural tendency to head toward the center of the universe while light things tended to move away from it. If the Sun were at the center, not the Earth, they would expect heavy things to naturally pull toward the Sun. This didn't happen. Rocks didn't fly into the air, heading for the Sun. And so the Earth had to be left in place, even if it meant building a byzantine structure to explain how planets moved.

The resultant fudged theory came to its height with the work of the second-century AD Greek astronomer Ptolemy. He put the planets on paths known as epicycles, effectively complex wheels within wheels that meant that instead of simply rotating with the crystal spheres the planets orbited a point that itself rotated around the Earth. This was a messy way of producing the same effect as would have been achieved by moving the Sun to the center of the universe, but it patched up the Earth-centered model and that was the description of the universe that would hold firm, with only slight deviations, until Galileo came on the scene.

We are used to thinking of Galileo, born in 1564, as a key figure in the debate over the heliocentric universe because of the famous trial caused by the book he wrote on the subject. But it was, of course, the Polish astronomer Nicolaus Copernicus whose book written in 1530, but not published until he was on his deathbed in 1543, who inspired the whole debate. Even so, Galileo was a major force in reinforcing Copernicus's theories, which did away with the need for epicycles, because Galileo backed up his theorizing with a whole swath of new astronomical observations.

Although Galileo did not invent the telescope—the essential vehicle for exploring the universe using light—he was the first to use it seriously to observe the heavens. Everything he saw seemed to drive a wedge into the ancient Greek description of the universe. For the Greeks, the Moon and everything above it had to be perfect, meaning that the Moon itself should be a smooth sphere—but the view through Galileo's telescope (feeble though it was) made it clear than the markings on the Moon's surface were not just patterns but craters and mountains that cast shadows. It was bumpy, not smooth. And then he pointed his small telescope, less powerful than a typical modern pair of binoculars, toward Jupiter. And what he saw persuaded him that Copernicus had got it right.

Galileo discovered four new "planets," which he realized were moons that orbited around Jupiter. With our modern orientation it is hard to see why this was so significant. Of course Jupiter has moons—like most planets. However, Aristotle and Ptolemy's universe could not encompass such a possibility because everything had to revolve around the Earth. If these moons orbited Jupiter instead, it made a lie of this one-time certainty.

The trial saw Galileo spend the rest of his life under house arrest, which could have resulted in his execution. It has been assumed that the threat of execution was because he dared to question the church's assertion that the Earth was at the center of everything. This certainly was church doctrine—both because there a couple of passages in the Bible that mention the Sun stopping in the sky (so by implication the Sun must move around the Earth) and also because Aristotle was considered an authority whose words were given nearly as much weight as those of the early Christian writers. However, it wasn't Galileo's astronomical theories that got him into so much trouble.

The book that Galileo wrote, *Dialogue Concerning the Two Chief World Systems*, had been okayed by the church authorities provided Galileo put in a disclaimer, effectively the words of the pope, making it clear that the Copernican universe was merely a thought experiment, not a true reflection of creation. In the book, Galileo featured three characters, who explore the ideas he covers in the form of a conversation. One of them, called Simplicio, was frankly not very bright. His role was to say, "Duh, I don't understand!" so the others could explain. Galileo's near-fatal error was to put the disclaimer in the voice of Simplicio, in effect casting the pope as an idiot.

THE UNIVERSE GROWS

If the church thought that banning Galileo's book (soon published elsewhere in Europe) and putting him on trial would suppress the heliocentric universe, it was mistaken. The simplicity of the theory, doing away with the messy epicycles, was far too powerful to resist for long. It made too much sense. From the seventeenth century onward, the accepted model of the solar system was, at its core, much like our own, but thanks to Galileo's successors in making use of telescopes, extra planets were added, and the universe itself grew in stature from the size of the solar system to take in the Milky Way Galaxy and eventually the current view of our living in a visible universe that is around 90 billion light-years across, though the full scale may extend much further.

Until the twentieth century it was visible light alone that helped us to explore the universe in space and time, taking in stars and galaxies and nebulae—but visible light is only a tiny proportion of the full spectrum of electromagnetic radiation available and flooding through the universe. Since the 1960s we have added radio, microwaves, infrared, ultraviolet, X-rays, and gamma rays to the astronomer's armory, opening up regions of the sky that were difficult to see or impossible to penetrate with visible light alone.

The biggest limitation presented by all these types of light is the barrier that came before the cosmic background radiation was let loose. Light simply can't penetrate back before 370,000 years into the early life of the universe. But in principle there is another mechanism to get even further back in time, because the universe is thought to have become transparent to a different type of particle, more elusive than photons of light, when it was only around 1 second old. This particle is called the neutrino.

SEEING WITH GHOSTS

It's no surprise that neutrinos can delve further into the past, through the opaque plasma of the superhot early universe, because these ghostly particles are very difficult to stop. They are produced in vast quantities by stars—around 50 trillion neutrinos hit your body every second from the Sun, for instance. But as far as the neutrinos are concerned, you are totally transparent. They fly through you as easily as they flitted through that early universe. Neutrinos are so bad at interacting with other particles that we have to really go out of our way to spot them.

It is telling that the existence of neutrinos was predicted in 1930, to explain energy that goes missing in fusion reactions, but it wasn't until 1956 that anyone managed to detect a neutrino when it made a very rare collision with another particle. A neutrino telescope is quite unlike any other. We tend to put our most powerful telescopes on mountains, or better still in space to avoid interference from earthbound sources, but neutrinos couldn't care less about interference. The detectors for them are typically placed at the bottom of deep mines, where pretty well all other particles are shielded out by vast quantities of earth and rock.

The "telescopes" themselves tend to be large containers of fluid—often simple cleaning fluid—surrounded by arrays of detectors that pick up the occasional particles that are produced in the body of the fluid. The vast majority of neutrinos pass straight through, but very infrequently one will interact with another particle, creating a tiny flash of light in the fluid. After eliminating local radiation sources, pretty well all that is left to cause this is the occasional interacting neutrino. This method has been used to produce a crude image of the Sun. As if to demonstrate just how little neutrinos care about matter, the Sun was at the far side of the Earth at the time. In the future, more advanced neutrino telescopes may give us the chance to peer back through that 370,000-year barrier and get a clearer idea of what really happened in those early years of the universe.

From the beautiful studies of distant galaxies and quasars produced by the Hubble Space Telescope, through the universe-spanning scope of the cosmic microwave background radiation provided by Planck, and from many other telescopic views on space from Earth and satellites we are constantly improving our light-based exploration of the universe. There is no doubt that there is far more to learn this way. And as we have seen, unless we develop some form of warp drive, this will remain our only way to explore the far reaches of the galaxy and beyond.

Yet ultimately, wonderful though the work of visual pioneers is, they will never truly fulfill the human urge to explore. It's a bit like watching a TV documentary about a distant land. It may well be fascinating and informative. It can provide all that we need to establish in the way of scientific data, but it will never be the same as actually *being* there. And so, despite all the risks and the cost, there will always be an urge to have a physical presence out among the stars.

There are a number of challenges in the way of such a feat. And the first, and in some ways the most formidable, is the warp in space and time that keeps us in place on the surface of the Earth. To explore space we first need to get out of the Earth's gravity well.

4.

ESCAPING THE WELL

||||||||||||||||||||||||||||||||||||||||||||||||||||||

Were it not for gravity one man might hurl another by a puff of his breath into the depths of space, beyond recall for all eternity.
—*Philosophiae Naturalis Theoria* (1758)
Roger Joseph Boscovich

Although the frontiersmen of America had plenty of challenges to face, they were, at least crossing the surface of a continuous continent, but their forefathers had to overcome a dangerous and difficult barrier first—venturing across an ocean in small, ill-equipped ships. Similarly, anyone trying to take on the final frontier of outer space has first to overcome another barrier: the gravity well of the Earth. Once you are free of Earth's gravity it is relatively easy to move around in space, but those first miles from the surface are the hardest.

KEEPING US IN PLACE

The gravitational pull of an object as massive as the Earth is considerable and there are two approaches that can be taken to breaking away from it, typified by Superman's baseball and a rocket. If someone throws something fast enough, whether it is a baseball hurled by Superman or a projectile, it will escape from the Earth's pull. Get it up to the breakaway speed—escape velocity—and nothing more is required. Once it is going that quickly, the acceleration of the gravitational pull that is slowing it down is insufficient to bring it to a halt and it will continue away from the Earth indefinitely. To go straight up, that velocity is around 11.2 kilometers per second (roughly 25,000 miles per hour)—quite a feat with a baseball unless you *are* Superman, but theoretically possible with the right technology.

It's possible to shave a little off this limit by getting help from the movement of the Earth. The Earth is, of course, spinning around, so something that is sent off with the direction of the Earth's rotation (eastward) picks up some extra speed from the Earth itself. Do this near the equator and you can get escape velocity down to around 10.7 kilometers per second (roughly 24,000 miles per hour). But this is still quite a challenge. By comparison, a rifle bullet travels at around 1,000 miles per hour. Although in principle a space cannon like Jules Verne's Columbiad could be produced, as mentioned in chapter 2, the g-force necessary to get a projectile up to the 10.7 kilometers a second mark before it leaves the barrel (because once it does there is no more force and all the spaceship can do is slow down) would turn a human being to jelly.

The only alternative, it seems, is to carry on applying a force to the projectile as it climbs. To do this the vessel has to have some form of propulsion built-in, but the huge advantage is that there is then no lower speed limit. A rocket could in principle leave the Earth at a walking pace, as long as it can keep up enough thrust to keep the vessel in motion at that speed. In general, the force applied is the mass of the object times the acceleration it undergoes as a result of that force. On the surface of the Earth, that acceleration is $9.81 \, \mathrm{ms^{-2}}$—around 32 feet per second per second. So for a human being with a mass of 75 kilograms (165 pounds) the force due to gravity—is around 735 newtons. Of itself it is difficult to get a feel for what this entails, so let's break it down.

The scientific unit of energy is joule. If something uses 100 watts, it uses 100 joules every second. And it takes 1 joule to move 1 meter against a force of 1 newton. (This is why scientists use metric units. It all fits together much more simply than trying to do the same thing in traditional units.) So to move 1 meter against 735 newtons would take 735 joules of energy. That's not a lot—the equivalent of running a 100-watt bulb for 7.333 seconds. But there are a whole lot of meters required. Say you want to get 100 kilometers (62 miles) up to be out into space. You might think you'd need $735 \times 100,000$ or 7.35×10^7 joules, or around 73 megajoules. That's the output of a typical power station for 1/10th of a second.

In reality things are a little messier. There is air resistance to overcome, to begin with. And the energy required will drop off as the gravitational attraction falls away when you are further from the Earth. Although the escape velocity is 11.2 kilometers per second at the surface, it is a little less as you get higher. It doesn't drop off as quickly as you might think, though. In the low Earth orbit of the ISS, for instance (around 415 kilometers or 258 miles up), the pull of gravity is still 90 percent of the value on the Earth's surface. The reason the astronauts float around is because they and the station are in free fall toward the Earth—only their sideways motion means they keep missing the Earth and continue in orbit. You would also need typically to move a lot more mass than just a human being to get into space. The Saturn V (pronounced "Saturn Five," not "Saturn vee") used to power Apollo into space, for instance, weighed around 2,800 tonnes (3,086 tons). So getting into space can take a fair amount of energy.

In looking at these numbers we have concentrated on the Earth's gravitational well—it is important to remember that the Sun has a much deeper well and to escape the solar system entirely from the Earth's surface, Superman's baseball would have to be thrown a huge 42.1 kilometers per second (94,175 miles per hour). To escape the Sun is a far more dramatic task than to get away from the Earth, though missions within the solar system rarely have to reach solar escape velocity as usually they can take a trajectory that isn't away from the Sun for the whole flight, or that uses the speed of a planet to give a slingshot increase in velocity.

Despite the fact that there is no need to get up to high speeds to head off out into deep space if you can apply thrust during the flight, you will often see something like this stated. In part this is because existing chemical rockets can only carry enough fuel to be under power for a relatively short time, so they have to reach Earth's escape velocity (at whatever altitude they are) before the burn finishes. But it is also because two separate requirements are being confused. It is quite true that a spaceship under power could make the journey from the Earth to the Moon, never exceeding a walking pace of 3 miles per hour. Admittedly it would take around ten years to get there, but it is possible. There is no need to reach escape velocity. But the reality is that most things we send up there don't head off into deep space.

The vast majority of the hardware that has been launched through the gravity well to date is heading for an Earth orbit. To achieve the kind of low Earth orbit used by the ISS and many satellites, a space vessel needs to be traveling at around 7.8 kilometers per second (17,500 miles per hour). That might be slower than escape velocity, but it still is a considerable speed, and it is this that I suspect many writers who insist that a ship needs to achieve escape velocity have in mind.

AN EQUAL AND OPPOSITE REACTION

Because it has been the power source of choice since the beginning of spaceflight, we tend to think of a rocket as the natural approach when it comes to driving a spaceship—whether to leave the Earth's gravity well or to head out through the depths of space. In the early days of rocketry there was some confusion over just what was happening at the back end, with the suggestion that a rocket works by pushing on the air (see, for instance, *The New York Times*'s mockery of Robert Goddard on page 50)—in fact air is a disadvantage for a rocket, which carries both its fuel and the oxygen required for that fuel to burn. All the air does is to provide resistance and slow the flight down. Rockets work perfectly well in a vacuum. Not only does a rocket not need that air for combustion, there is no need for it to have anything to push against because rockets are devices that are set in motion by Newton's third law.

You may remember Newton's laws of motion from high school. They are simple yet elegant statements of the way moving things behave, built on Galileo's work. When not peering at the skies, Galileo transformed physics, throwing out centuries of acceptance of the ancient Greek view that things had a natural tendency to stop and you had to push them to keep them moving. Galileo came up with, and Newton encapsulated in his laws, the idea of inertia. They realized that instead, things keep moving the same way (or keep on *not* moving) unless you do something to influence them.

This wasn't obvious to the Greeks because in their world of wood and stone devices, with little in the way of precision smooth-moving metal parts, friction was king. Friction and resistance to movement seemed so much a natural part of existence that it didn't strike them that this was a result of something acting on moving objects, rather than their natural tendency. This idea of inertia sets up Newton's first law that something will continue in uniform motion unless a force acts on it, while the second law describes how much force it takes to speed something up or slow it down. But the third law is the key to the mystery of the rocket.

You can demonstrate this law in action yourself if you stand on ice skates, or balance on a low friction trolley. (*Warning:* This experiment puts you at serious risk of falling over. Try it at your own risk.) If you have with you a heavy weight and throw it away from you, you will start moving in the opposite direction to the weight. This has to happen because momentum—the mass of an object times its velocity—is one of the quantities that is conserved in any particular system. So if you give the weight some momentum in one direction, you must gain momentum in the opposite direction to balance it out. Every action, Newton, made clear has an equal and opposite reaction. Force an action on the weight—the result is an equal and opposite reaction on you.

So this would seem to suggest rockets should work by throwing weights out the back—and in a sense they do. Rockets have been so much part of our lives for so long—far longer, for instance, than the internal combustion engine—that it can be easy to take an understanding of them for granted. But it really is rocket science, so it's worth taking a moment to think a little more about how they work. Technically one of the earliest known examples of this kind of action was Hero of Alexandria's aeolipile. This was a little metal ball with water inside. A pair of bent tubes emerged from out of the sides of the ball. When the water was heated, steam pushed out of the tubes, which, when pointed in the right directions (opposite each other), cause the ball to rotate. The aeolipile was a crude steam rocket. But it's not really what we usually think of when we hear the term.

FROM FIREWORKS TO TERROR WEAPONS

The familiar chemical rocket had its origins in China, where gunpowder-propelled rockets were already in use for both entertainment and as weapons of war by the twelfth century. We know that knowledge of gunpowder had made it as far as Europe by about AD 1270, because details of it are mentioned in a remarkable letter written by the English friar Roger Bacon. (Despite the myth, there is no evidence to suggest that Bacon invented gunpowder. It was almost certainly brought west from China.) Bacon's letter describes a rather poor recipe for making black powder or gunpowder, one better suited to producing flashes and bangs than giving good propulsion to a rocket (or a bullet) but once the black powder cat was let out of the bag, there was no going back.

In chapter 2 we spent some time on the possibility, popular in early science fiction, of using giant cannons to propel craft into space—the process of firing a gun and a rocket use the same basic principle, but in a back-to-front fashion. In a gun, the gunpowder is ignited in the barrel. As the gases produced by the rapid combustion expand, they force the bullet along the barrel until it exits. Once the bullet is out of the barrel, it is in free flight. A bullet truly would have to reach escape velocity to get into space. With a rocket, though, the fuel stays with the projectile and can continue applying force to the ship long after it leaves the surface of the Earth.

When a chemical rocket, using gunpowder or any other combustible fuel, burns, it is just like throwing rocks or pieces of iron from the back end, but instead it is the molecules of the exhaust gases that blast out of the rear of the rocket. Bear in mind what is being conserved is momentum, mass times velocity. Those gases will weigh a lot less than your rocket—but they are moving at a much higher velocity, and have plenty of thrust because they are accelerated extremely quickly by the near-explosive burn of the fuel. The relatively small mass of the gas, blasting away at high speed makes the relatively high mass of the rocket move in the opposite direction at a slower speed—but it is enough to get it moving. And the longer you apply a force (especially as the air resistance falls as the atmosphere thins), the more speed you end up with.

Rockets began to be used seriously for military applications in the West from the nineteenth century, though at this stage what was involved was little more than a large-scale version of a bottle rocket with a metal casing rather than a cardboard one, carrying an explosive charge. This is the kind of thing that Francis Scott Key had in mind in his "Star-Spangled Banner" when he wrote of the "rockets' red glare." The source of these rockets was the British vessel HMS *Erebus,* which was converted to fire so-called Congreve rockets, weapons that had been inspired by the rockets used against the British in India from the late eighteenth century.

Rockets were frequently deployed from ships because, unlike artillery, firing a rocket does not produce a recoil. (This is also why a rocket-propelled grenade can be fired from a launcher held on the shoulder. A similar sized gun would take the shoulder off with the power of the recoil.) The action and reaction is between the rocket and its exhaust, leaving the ship unaffected. The Congreve rockets were impossible to aim with any accuracy, so acted more as a weapon of terror than an accurate piece of artillery. To have any sensible control required a more governable fuel and far more sophisticated steering technology.

RUSSIAN IDEAS, AMERICAN EXPERTISE, AND GERMAN VISION

At a conceptual level, the basics of modern rocket science came from Russia, where high school teacher Konstantin Tsiolkovsky worked on the underlying theory and suggested that liquid hydrogen and oxygen would make a far more efficient and flexible fuel than black powder. Tsiolkovsky's masterpiece on the subject *Investigation of Outer Space Rocket Devices* was published in 1903, the same year as the Wright brothers took to the sky for the first time.

Among the firsts that Tsiolkovsky devised was the idea of mixing fuel and oxidizer in a reaction chamber, rather than simply igniting the whole thing, as happens with solid fuel, and the realization that it would be necessary to have multiple-stage or "step" rockets to get a large mass into orbit. The idea itself was not new, but Tsiolkovsky was first to appreciate how essential it would be for manned flight. By dropping the heavy bottom stages once their fuel is used up, there is less mass to be propelled later in the flight, making the rocket far more efficient.

This driven Russian schoolteacher was a theorist rather than a practical rocket scientist. The hands-on side of early rocket development came from America, where Robert Goddard experimented with liquid fuel rockets and made the first practical trials of giving rockets multiple stages to achieve greater range. Goddard's rockets were primarily built on a small scale, but his vision was massive. In 1920 the Smithsonian Institution published his paper *A Method of Reaching Extreme Altitudes*, where he dared to suggest that a rocket set off from the Earth could reach the Moon. The press went wild.

Looking back from the present day, when Goddard is very much regarded as *the* outstanding pioneer of practical rocketry it is something of a surprise that the press response was not one of universal adulation, but of scorn. Referring to him sarcastically as the "moon man," the press simply could not believe that Goddard had his facts right. *The New York Times* infamously chided Goddard for his silliness in an editorial remarking, "That Professor Goddard, with his 'chair' in Clark College and the countenancing of the Smithsonian Institution, does not know the relation of action to reaction, and of the need to have something better than a vacuum against which to react—to say that would be absurd. Of course he only seems to lack the knowledge ladled out daily in high schools."

The editorial writer had clearly got his Newton's laws in a twist, believing that a rocket flew because the exhaust gas pressed against the atmosphere, rather than that the force exerted on the exhaust produced an equal and opposite push on the rocket itself—which is what actually happens. The newspaper did make a "correction" to its insulting editorial in 1969, when *Apollo 11* was in flight, commenting that further investigation and experimentation have confirmed the findings of Isaac Newton, though it is a shame that this wasn't obvious to the newspaper's writers forty-nine years earlier.

Goddard was probably horrified by the response he received, and managed his relationship with the media much more carefully after this shameful treatment, being careful to avoid unnecessary mention of manned spaceflight, which in hindsight was probably a pity in a field that is so driven by public attitude and politics. He went on to perform many experiments in his new site, ironically located, given its later infamy as the conspiracy theorists' favorite home of crashed UFOs, at Roswell, New Mexico. There he got a rocket up to an altitude of eight thousand feet, built a rocket that traveled faster than sound, and investigated a range of inertial guidance systems that used gyroscopes to control the direction of flight, an essential if rockets were to get past the randomness of the bottle rocket's flight.

One final engineering dreamer of space travel deserves a mention at this stage—the German scientist Hermann Oberth, who Fritz Lang and George Pal consulted for their groundbreaking movies. Unlike Goddard, there was no coyness for Oberth about the reasons that he was involved in rocketry. Oberth would have regarded being called a moon man a compliment. His breakthrough work, published in 1923, was entitled *Die Rakete zu den Planetenräumen* (*The Rocket to Interplanetary Space*). The book included impressively detailed plans for a liquid-fueled, two-stage rocket incorporating a spacecraft that was intended to carry human beings away from the Earth.

Like Tsiolkovsky, Goddard and Oberth were driven purely by an enthusiasm for the romantic possibility of space travel. They had been inspired by reading the likes of H. G. Wells and were arguably dreamers detached from the real world. But the kind of rocket that would be used to get satellites and human beings into space instead had a genealogy that took the rocket back to its earlier role of a terror weapon. We stay in Germany, not in Oberth's utopian world of mankind traveling into space and opening new frontiers, but rather in the Germany of the Third Reich. The next big step in the journey into space came from the rocket's return to warfare as a terror weapon of the Second World War.

FROM NAZI TO NASA

German engineer Wernher von Braun, who was strongly inspired by Oberth's writing and briefly worked with the older man, had been working on a powerful missile since before the beginning of the war. But this was very early in the development of liquid-fueled rockets, and many of his smaller prototypes exploded or crashed. Pressure remained on him to succeed, though. The German hierarchy wanted something better than the "doodlebug," the V-1 terror weapon (the "V" in the code name stood for *Vergeltungswaffe*— revenge weapon) that was powered by a crude pulsed jet engine, acting like a primitive cruise missile.

The V-1's successor was to be provided by von Braun's A-4 rocket, renamed the V-2, a much bigger, true rocket that was the first ballistic missile, one that is powered up into space, then glides in free flight before plummeting back down into the atmosphere and toward its target. First used on Paris in September 1944, these forty-six-foot-high missiles were particularly feared in their main targets, London and Antwerp, as they seemed to appear from nowhere, plunging out of the sky so quickly to deposit around a ton of explosive, that they were not noticed until the explosion came.

Perhaps even more horrific than being in a city on the receiving end of the V-2 was living through the working conditions of the concentration camp where forced labor was compelled to work in the new underground home of von Braun's A-4 production after the original site was bombed. Albert Speer, the German minister of armaments, no stranger to the Third Reich's atrocities, commented years later about a visit to von Braun's factory: "It was the worst place I had ever seen. Even now when I think of it, I feel ill."

Given his later huge contribution to the U.S. space program, there are mixed feelings about von Braun's involvement in the Nazi regime. He was certainly a member of the Nazi party and became a member of the feared SS (though quite possibly this was under pressure). On the other hand, von Braun was briefly arrested by the Gestapo for putting his ideas of space travel above the needs of the military. It seems likely indeed that von Braun had no real interest in the war, but merely saw his work on missile development as a means to work toward his space travel dreams. Even so, from a modern viewpoint it is impossible to forget the nightmare environment where the concentration camp workers who made his vision possible lived and died. Around ten thousand prisoners lost their lives working on von Braun's project—twice as many as were killed in allied countries by the V-2 attacks.

The liquid-fueled engines, with liquid oxygen on tap to keep the fuel burning out of the atmosphere, and the tall, dart-like shape of the A-4, launched vertically, would become the prototypes for all future space rockets. In part, this was because the A-4 project was so successful, the first practical large-scale rockets to fly reliably, and, in part, because unused V-2s and their associated equipment (and staff) were captured by both the American and Russian armies at the end of the Second World War. The American space program gained a huge boost in that von Braun also found his way to America and would make a major contribution to NASA's early work.

THE FIRST PASSENGERS

Although the Russian dog Laika would be the first animal to make it into orbit (see page 72), early U.S. experimental work with captured German A-4 rockets did include a number of flights scraping the edges of space with monkeys as passengers, used with the clear intention of preparing for a future when rockets would carry human passengers. Unfortunately most of the missions, code-named Albert after the first of the monkeys, were not successful. They all intended to bring the monkey back alive, but all six of the first monkey subjects died, mostly due to parachute problems that resulted in crash landing. (Sadly, though Albert VI survived landing, he was left in the capsule for too long and died of heat exhaustion).

It was only on the seventh attempt in 1952, using new, smaller, U.S.–designed Aerobee rockets brought into use when the captured A-4 stock was used up, that a double parachute system succeeded in bringing back a pair of monkeys safely. The Russians too had early animal experiments, though they had more success with their dogs, first getting a pair up to 63 miles above the surface and returning them alive in 1951.

IT'S ONLY ROCKET SCIENCE

Most of the development of U.S. rocket technology would take place at the Jet Propulsion Laboratory in Pasadena, California (technically it is in La Cañada, California, but this city did not exist when the JPL was first set up). The name of the facility is ironic, as its entire life this has been a site to work on rocket propulsion, not on jet engines. According to JPL legend this is because when the facility was first set up in the 1930s, rockets were still considered far too much the fodder of science fiction to have a serious U.S. base dedicated to them. Some engineers refused to work with them at all, and the rocketry had to be concealed by the less controversial "jet" term. (The same goes for the rockets used to give aircraft short takeoff runs, known as JATO or jet-assisted takeoff, even though jets have nothing to do with it.)

Until recently, most rocket technology has been very much a spin-off of the military requirement for launchers to carry long-range intercontinental ballistic missiles (ICBMs). So, for instance, the Atlas rockets initially designed to carry ICBMs would be responsible for getting the first American astronauts into orbit. Although other launchers, like the Saturn V used for Apollo, were never designed for carrying weapons, they could still trace their lineage back through the Jupiter series to the Redstone missile and hence back to the V-2 missiles. After all, the hand behind the Saturn V was that same Wernher von Braun who masterminded the German weapon program.

The big problem with taking the rocket approach to getting away from the Earth is that you need to carry your fuel with you. This might not seem much of a problem if all you want to do is drive a car down the highway, but the mass of the fuel is already an issue for an airliner. Aircraft are fueled with kerosene, which like gasoline is a mix of hydrocarbons refined from crude oil. Aviation fuel has the useful trait of burning efficiently and producing a considerable amount of energy per unit weight. Aircraft fuel typically has larger molecules than the petrol or diesel used for cars, and is less volatile.

The big advantage of aviation fuel over, say, electric batteries, is that it really packs in the energy, meaning that the plane is able to carry the weight of fuel on board. To have the same energy content as a ton of batteries would only take around 10 kilograms (22 pounds) of aviation fuel. Even with this ability to cram the energy in, a 747 carries up to 60,000 gallons of fuel which weights 186 tonnes (205 tons), 42 percent of its entire maximum takeoff weight. So a 747 burns a fair amount of fuel just to keep its fuel in the air.

For a rocket things are far worse. To carry enough fuel to escape the Earth's gravity well becomes a real issue of packing in enough energy, without the increase in mass from the fuel itself requiring so much extra energy that it will never succeed. The most familiar solution to this issue, as we have seen, is staging. When you watch a video of a Saturn V rocket carrying an Apollo capsule away from the Earth there were three separate stages of the rocket, each with its own motors. The point of this exercise is that the extra stages give more capacity for fuel, but as soon as a particular stage is exhausted, the whole section including all the mass of metal, is discarded, leaving much less mass for the next stage to have to accelerate.

HARVESTING OXYGEN AND ELECTRIC ROCKETS

Traditional staging is not the only way to get around the weight of fuel a rocket needs to carry. Commercial space venture Virgin Galactic (see page 91) intends to use an aircraft as a first stage of the flight to get the rocket propelled vehicle as high as possible before needing to eat into the available mass of rocket fuel. An alternative tactic is the innovative approach taken by the proposed Skylon spaceplane using the SABRE propulsion system developed by Reaction Engines. Skylon, a space delivery vehicle that takes off and lands as a conventional plane, uses liquid hydrogen as its fuel like many rockets. But while it is in the atmosphere, it takes in the oxygen needed to combust the hydrogen from the air, meaning it needs carry significantly less fuel (or technically less oxidizer for the fuel).

This might seem an incredibly obvious step to take. Why use stored oxygen when you are flying through a free bath of the stuff? Yet current hydrogen/oxygen rocket motors, like those used on many NASA missions, all carry their own liquid oxygen. This is because there is a huge technical challenge to overcome. To work with the hydrogen fuel in a rocket, oxygen has to be compressed and cooled down to around −140 degrees Celsius (−220 degrees Fahrenheit) before it is mixed with the hydrogen for combustion. This process has to take place in around 1/100th of a second, while overcoming the problem of the moisture freezing out of the atmosphere and clogging the piping with ice. Reaction Engines has successfully demonstrated a special heat exchanger featuring a helium-cooling loop that successfully did just this, and the hope is now to have a fully working SABRE engine by around 2015.

Reducing the need to carry oxidant is a first step, but along the way we will meet a whole range of mechanisms to propel a spacecraft. As yet, rocket-based motors are around the only way we have of getting out of the Earth's gravity well, but once the ship is out there in open space, ion thrusters have already proved popular. These may sound like pure science fiction, but ion thrusters are a kind of electric-propelled rocket, where charged particles (ions) produced using heat or radiation are flung out of the rear of the craft using electrical repulsion to accelerate them.

As the ions are pushed out of the back of the craft, Newton's third law then provides an equal and opposite thrust, powering the ship through space. Ion thrusters don't have enough power to get a ship out of the Earth's gravity well because they can't produce as much force as the pull of gravity on the Earth's surface, but they can be made very small, so are ideal for use as maneuvering thrusters, and are liable to have a major role in deep-space flight.

To make ion thrusters work you need to carry two substances—the atoms that will become ions and the power source for the electricity, though the latter can, to some extent be derived from solar power. However, other modes of power available once in space are more frugal. The most dramatic examples of fuel-free flight come from solar sails, using the thrust from the light and streams of particles emanating from the Sun. These have been demonstrated to work, but have not been truly used yet. They provide a very low level, but steady thrust, while the craft is in the relatively near vicinity of the Sun. Other designs look to pick up fuel, or the reaction mass for an ion thruster, from space itself. Space may be mostly empty, but there are plenty of atoms of gas and fragments of dust floating out there that could be scooped up as a ship flies. As yet, most such designs would need a lot of work to make them even potentially usable.

ENTER THE ATOM

The dark hero (or, some would say, antihero) of space propulsion is the use of nuclear power. Although we are naturally wary of making use of nuclear power on the Earth, there are many ways that it can be deployed to power a spaceship, which could benefit hugely from the much higher energy production for any particular mass of fuel than is possible with a chemical rocket. A nuclear generator can be used as the power source for electricity with ion thrusters, or it can be used to heat a liquid or gas, much as happens in a nuclear power station, which can then be sprayed from the rear of the craft to produce a more conventional rocket motor effect.

If nuclear fusion can be used, then the motor could use the actual material being fused as its reaction mass, turning its fuel into the stuff that is thrown out of the back of the rocket. This is, however, a big "if" given the slowness of advances in nuclear fusion techniques for power stations, which have been under development for over fifty years and are still to get to the position where a fusion generator produces more energy than is put in. But that is only the start of the possibilities for nuclear propulsion, which range all the way to the apparently outrageous suggestion of setting off a string of nuclear bombs behind a craft and riding the shockwave like a surfer. More on this approach later.

The difficulty of escaping Earth's gravity well with conventional rocket motors is one of the two reasons for the enormous cost of getting any mass into space. The other issue is that there is a lot of extra baggage to be carried to keep the human payload alive. NASA currently estimates that the cost is around $20,000 per kilogram ($10,000 per pound), while commercial rockets have got this down to around $5,000 per kilogram ($2,200 per pound). So that's between $375,000 and $1.5 million for a typical person, while the overall cost including the ship and supplies could easily take the figures twenty times higher. This is never going to be an everyday kind of activity, however good your rocket.

The cost of getting into space is an economic iron first restraining our exploration that has to be shattered if we are ever to expand to the final frontier in a big way. What has been consistently true of Earth's pioneers is that while it might be expensive for the very first people to get to the frontier, opening up and making anything significant of the opportunities it presents requires a way to get people out there cheaply. As yet we don't have a way to do this for space, and though the competitive pressures of commercial spaceflight are bringing the costs down compared with the first attempts, to make all the attractive possibilities from space mining to true exploration possible it is essential that we get past this first hurdle.

One possibility is to avoid the Earth's gravity well as much as possible. As we will see later on, building a mining base on the Moon is in some ways a more attractive option than shipping materials from the Earth because a lot of the raw materials could be obtained there and either made into parts on the lunar surface, or dispatched into space. The Moon's lower gravity makes it significantly easier to get away from and makes practical the use of mechanisms that may never work on the Earth like a space elevator (see page 61) or the dramatic-sounding mass drivers.

CATAPULTING INTO SPACE

A mass driver uses a series of electromagnets to accelerate a payload along a track to such a speed that it can be catapulted into space—it is Superman's baseball writ large. In essence what is required is something similar to the linear accelerators used to push particles to huge speeds, but applied instead to a large metal container. From the surface of the Earth, the acceleration required and the length of the track necessary to get up to speed render the idea impractical, but, say, on the Moon, providing the challenges of undertaking a major engineering project in the harsh lunar environment could be overcome, the idea would be entirely practical.

To escape the Moon without subsequent thrust being applied, a ship would have to be accelerated to around 2.3 kilometers per second (5,219 miles per hour). This sounds a lot, but it could be achieved with a 27-kilometer (16.8-mile) long track, providing an acceleration of 10 g—just about acceptable for humans. With robust inanimate cargo, not at risk from the g-force, the track could be shortened to one-tenth that length, provided the driver could produce the 100 g acceleration required. If such a device were built on the Moon, the idea of mining rare materials (see page 169) becomes much more cost effective than if it has to be blasted off in a rocket. Although the setup cost of a mass driver is large, once in place, the day-to-day cost of getting a capsule of mined materials off the surface is tiny compared with using chemical rockets.

A more off-the-wall related approach is the so-called space fountain. This uses some form of ground-based accelerator, like a mass driver, to push vast numbers of small pellets up to high speed. These pellets shoot up into the air, eventually falling back down. Because they return to Earth (or the Moon) with a considerable amount of kinetic energy, this energy can then be harnessed in reaccelerating back up into space, meaning that it takes very little energy to keep the pellets in motion once they are first brought up to speed. The idea is then to propel a craft into orbit by bouncing these pellets off the bottom of the craft. This has been demonstrated to work on a small scale, but would be a huge challenge to scale up. Keeping a craft stable is a delicate balance, and it seems unlikely this would ever to be used for a full scale launch, but it remains an interesting concept.

ELEVATOR TO THE STARS

The alternative, whether on the Earth or on the Moon, is to find a different way to get up there that does not require vast amounts of fuel, and that avoids the need to carry that fuel (and its mass) in the vehicle. There is one possibility more practical that the space fountain for doing this—the so-called space elevator or skyhook. While this has a major technical issue to overcome, particularly for use on Earth, it could transform the interface between our planet and space, making transportation of people and materials to space bases and construction of deep-space ships trivial.

The basic idea is so simple and elegant that it can't fail to appeal. It's what you might call a Rapunzel technique. Just as the fairy-tale character let down her hair from the top of a high tower so her rescuer could climb up, with a space elevator, a satellite lets down a cable (let's not worry too much what that cable is for the moment) all the way to the ground and some form of elevator hauls its way up the cable to the satellite. The great thing is that this "crawler" can be a simple electrically powered device, with the power fed through the cable or, if this proves impracticable because of transmission losses, beamed up via a laser. There is no need for dangerous rockets or for the crawler to carry tons of highly flammable fuel. The result is, once the whole elevator system is in place, to make the cost of getting mass into orbit comparable to flying it in a conventional aircraft—the cheaper the electricity generation method, the less it will cost.

The (literal) make-or-break aspect of a space elevator is the cable. This would need to be constructed from a material that is both extremely strong and extremely light—because it is going to be very long indeed. To pass a cable between a satellite and the Earth, the satellite needs to be geostationary—fixed in position above the Earth's surface—or the cable will soon snap as the satellite heads around the Earth. Unless the satellite is constantly under power to keep it in position—a tricky option at best—there is just one height at which a geostationary satellite can stay in orbit. This is because an orbit is a delicate balance between falling and moving sideways to miss the Earth. Orbit too quickly for any particular height and your satellite will fly away. Orbit too slowly and the falling part wins and it crashes to Earth.

As the satellite needs to go at the same rotational speed as the Earth to be geostationary, there is only one altitude at which it can remain fixed in the sky, and that is 35,786 kilometers (22,236 miles)—which calls for a pretty long cable. Compare it, for example, with a cable that went around the entire Earth, which would only be a little longer at 40,000 kilometers (24,850 miles).

In practice a real space elevator cable could not simply run down from a geostationary satellite because the combined system with the cable would have a center of mass well below the geostationary height. Instead it would need a counterweight beyond the geostationary level, sufficiently high that the center of mass is *above* the geostationary level and hence keeps the cable taught. In the calculations below I am using the geostationary height as an example cable length, but in practice the cable would have to be significantly longer.

Imagine that the elevator was based on a wire rope (steel cable) 28 millimeters or just over an inch across, capable of holding a weight of around 50 tons. The cable itself would weigh around 115,000 tons. So not only would there be the difficulty of getting this enormous mass of material into space, the cable would not be strong enough to support its own weight and would fall apart. The mass problems would be even worse for the original concept of the space elevator, dating all the way back to the 1890s when Russian engineer Konstantin Tsiolkovsky, reputedly inspired by the newly built Eiffel Tower in Paris, suggested building a tower all the way to the geostationary height.

GROW YOUR OWN ELEVATOR

Perhaps the best hope at the moment for a practical space elevator above the Earth is to use carbon nanotubes. We are most familiar with two structures of carbon—graphite, which comes in atom-thick sheets that easily slide over each other, making it very soft and easy to apply (for example in a pencil lead) and diamond, which has an intensely strong three-dimensional lattice. But it is also possible to have the equivalent of a layer of graphite that is curled around into a tube. These structures are very light and have the highest tensile strength—resistance to being pulled apart lengthways, say by their own weight—of any known material. A cable 1 millimeter (0.039 inches) across can support over 6 tons.

At the moment carbon nanotubes (which are typically grown like a crystal rather than manufactured in the traditional sense) only come in extremely small lengths and large quantities of the tubes are used, typically bound into a medium to act as reinforcement. But if they could be produced at considerable length they could be made into a cable that was both very light and unmatchably strong, perhaps strong enough to cope with an earthbound space elevator.

One encouraging step forward is the establishment of Elevator:2010, a prize contest similar to the Ansari X Prize (see page 91), which has the backing of NASA and that has set up a number of challenges relating to the problems faced with building a space elevator. These include construction of the tether (or cable) and a beam power challenge to provide means of getting energy to a climber. The first level of this (with a prize pot of $900,000) was won in 2009 by LaserMotive, a company that used lasers to provide power to a 4.8-kilogram (11-pound) climber, which worked its way up a 900-meter (nearly 3,000-foot) cable to a helicopter.

For the moment most effort is focused on climbers simply because it is not clear how to get a strong enough cable—but developments are constantly underway with nanomaterials and it is certainly not inconceivable that such a material could be developed one day. Some argue, though, that it may never happen. While it is dangerous to say "never" in terms of development of materials, there are plenty of other problems facing an Earth-based elevator including overcoming vibrations in the cable, and the danger that could be posed by a 22,000 kilometer (13,700 mile) cable snapping (either accidentally or under terrorist attack) and whipping its way across the Earth's surface.

Whether or not a space elevator is ever built to escape from the Earth's gravitational field, if the reasonable idea of using the Moon as a (relatively) low gravity source of raw materials took hold, the position would be quite different there. Because the Moon's gravitational pull is only around one-sixth of that on the Earth, the cable for a space elevator constructed there would be much shorter and less heavy. So much so that conventional materials—particularly something like Kevlar, technically polyparaphenylene terephthalamide, the very strong fiber used in ballistic vests and other defensive garments, has a high enough strength to weight ratio to be able to work in a Moon-based space elevator.

Though establishing a manufacturing base and energy source on the Moon would still require a large quantity of mass to be taken from the Earth for initial construction (though the raw material for manufacturing are mostly in place on the Moon), once the site was set up there is the opportunity to use a space elevator to carry both the products of mining and parts constructed for long-range spaceships out into space at much lower cost than is possible using rockets. Most of the other issues with an elevator on the Earth—what would happen if the cable snapped or was attacked by terrorists, for instance—present significantly fewer problems when constructing an elevator on the Moon.

The Moon could also support a different kind of space elevator— more a space paternoster, the unusual style of elevator occasionally seen in buildings that has a series of cars constantly in motion, which the passengers hop on and off as it passes slowly by. (The same approach is also taken on some Ferris wheel rides, like the London Eye.) The rotating space elevator would pivot around a point that was in low orbit around the equator. The mechanism would be designed to swing at just the right speed to match the rotation of the Moon. It would come back to the surface at six points around the equator providing both a low cost way of getting into space and a low energy transit system for getting around the lunar surface, one-sixth of the Moon's circumference at a time.

BEAM ME UP, SCOTTY

In the end, though, elevators are likely to be slow—they would probably take several days to reach the point where cargo was detached and are limited in the amount of material they can carry in one go. As an alternative, it might be tempting to wonder why no one is working on a real-life version of the transporter used on *Star Trek*. The fictional version was devised both to save money (because it did not require the expensive model work that filming a shuttle launch and landing needed) and to save time, because landing in a ship is a relatively time-consuming process, leaving a hiatus in the forward movement of the drama. Oddly enough, in real life, if a transporter were possible to build it would probably take far longer than a rocket or even space elevator to get material from A to B.

There is a real scientific process called quantum teleportation that is, in effect, a *Star Trek* transporter on an incredibly small scale. Teleportation, which uses the strange phenomenon of quantum entanglement, overcomes a fundamental issue with duplicating a quantum particle like the atoms that make up a human body or a part of a spaceship. It is impossible to discover the exact state of a quantum particle, because the act of measurement changes the particle. This means you can't choose to put a duplicate particle into exactly the same state as the original, a process would be needed to build a transporter, using conventional means. But entanglement provides a way to get around this.

While you can't find out the properties of the particle you want to teleport, what you can do is to transfer those properties to another particle without ever finding out what they are. This is what quantum teleportation entails. I might start with a hydrogen atom, for instance. Using entanglement, I can take another hydrogen atom at a remote location and send a stream of information and particles to the remote location that will turn the second hydrogen atom into a perfect duplicate of the first. In the process, the first atom will become messed up. It isn't destroyed, but it will no longer be in its initial state. Anything made of such atoms would effectively be disintegrated.

What quantum teleportation really provides is a sort of remote matter duplicator that would destroy the original object in the process, leaving behind a collection of scrambled atoms. To make it work to transfer objects into space you would need appropriate atoms already at the destination. The teleporter doesn't create atoms from energy or send them from place to place, all it sends is information. The process then imposes a set of properties on matter that is already at the destination to convert it into the correct state. So, for example, to send a steel piece of a spaceship into orbit you would need a collection of iron and carbon atoms out there all ready to use to assemble it from.

Transferring the information, however, isn't the biggest of the problems you would face. To work on any macrosized object, be it a spaceship part or a human being, your transporter would have to scan every atom in that object. Let's say you used such a device on a human body. There are a vast number of atoms in your body—in round figures, 7×10^{27} (where 10^{27} is 1 with 27 zeros following it). There is no currently imaginable way to scan all those atoms at once. Imagine you had some kind of process that could strip of a layer of atoms at a time, perhaps working through a trillion in a second. That sounds impressively quick. But it would still take around 200 million years to scan your whole body. And then you would have to reverse the process at the other end.

The other problem from a human viewpoint is that this is not a genuine means of transport. It is a mechanism to strip your body into its component parts, scan them and assemble a perfect duplicate. Even if it were practically possible, the "you" that entered the transporter would die horribly. To the rest of the world you appear to have transported. The new you would be physically identical, down to your every thought and memory. But it doesn't sound much fun for your original version.

In practice then, quantum teleportation is a nonstarter for living things, even if it were possible to overcome the limitations of scanning and assembly. Although there are no ethical objections to taking this approach with a spaceship part, it is a ludicrously over-engineered way to get an inanimate object into space. What is much more likely is that some form of 3D printing would be used to make a close enough duplicate for engineering purposes, without worrying about getting the quantum state of the atoms right. But even then, the raw materials for the 3D printer (or for the teleporter) to use still need to be got into orbit. These are mechanisms for transferring a design, not mass into space.

DEEP-SPACE DESIGN

Once you get entirely outside the Earth's gravity well (technically, because gravity doesn't stop, you never leave the well, but what's really meant is when the impact becomes negligible), a whole new world of spaceship design opens up. To get away from the Earth (or other planets) requires a ship's structure that is extremely strong— hence heavy and requiring a huge amount of energy to escape the well. What's more, for the portion of that escape that involves traveling through the atmosphere, the ship needs to be aerodynamic. All those considerations go out of the window for a ship that is built in space and is never intended to land on a planet. Movie and TV science fiction often gets this wrong.

The USS *Enterprise* in *Star Trek*, for instance, is far too often portrayed entering the atmosphere or even, in the 2013 movie, not only managing an atmosphere and gravity well, but happily submerging under the sea and reappearing in one piece. No doubt Star Trek fans will explain this all thanks to some kind of force field, but in the real world, a ship of that size and weight would have to stay in space and rely on shuttles (or transporters) to reach the surface. This seems to have been understood in the TV shows, but has rarely been observed in the movie versions of the franchise.

A much more realistic image of an interplanetary ship, built in space and never intended to come near a planet's surface as we've already seen is the *Discovery*, voyaging to Jupiter's moons in 2001: *A Space Odyssey*. This was designed as a long, ragtag collection of modules that made no pretention of being aerodynamic. The appearance of the ship reflects *2001* author Arthur C. Clarke's observations in a nonfiction book on space travel written in 1951. Clarke pointed out that a deep-space ship would not need streamlining— the shape could be determined by engineering requirements alone. What's more, such a ship, undergoing long but low acceleration does not need massive strength to cope with its mass. He likened the structural integrity of such a ship to a Chinese lantern, commenting "and perhaps the analogy is not a bad one as the tanks could, at least for some fuels, be little more than stiffened paper bags!"

Before such a ship could be employed, though, we still need to get off of the Earth, and in managing this, for the moment, the picture remains fairly bleak if we are to open up the frontier of space. This doesn't mean that nothing is possible—nor that nothing has been done. We can chip away at the costs of getting into orbit or deep space with conventional rockets. But we can't expect huge reductions in less than decades.

In the meantime, though, we have got some experience. We may not have strayed much from our own backyard, but we had at least have practice at getting things into orbit—and life on Earth has benefited hugely as a result.

5.

BACKYARD EXPLORERS

||||||||||||||||||||||||||||||||||||||||||||||||||||||||||

The significance of the Soviet accomplishment in exploring outer space has been considered at length by the Board of the National Science Foundation. The Board regarded this as a great scientific and technical achievement; and urged that it be recognized as such.
—Statement regarding the Russian Satellite (1957)
U.S. National Science Board

The satellite is, of course, most widely and readily accepted as proof of scientific and technical leadership by those with the least scientific and political sophistication. The degree to which informed scientific and political opinion believes that the USSR has surpassed the US in scientific capability cannot yet be assessed. Sophisticated opinion is, of course, far less likely to be impressed by the drama of the satellite or its being a "first."
—Reaction to the Soviet Satellite—a Preliminary
Evaluation (1957)
White House Office of the Staff Research Group

Practically all our attempts to get out into space to date have been the equivalent of taking a trip around our own backyard or enjoying a brief boat trip around the bay. This doesn't make such exercises worthless—after all, most of the direct, day-to-day benefit we currently get from space is from Earth-facing satellites bringing us weather or pinpointing our locations with GPS. But this is a trivial step in terms of space exploration. Even so, it is all the foundation we have, and we need to put our attempts in context before we can look further.

SPUTNIK THROWS DOWN THE GAUNTLET

When looking back over our space achievements to date, so much emphasis is put on the Apollo program (for obvious reasons) that it might seem easy to forget that humanity's first venture into space travel involved no human beings, was not an American enterprise, did little more than broadcast a radio signal, and dated back to the antiquated technology of the 1950s. Yet for those who were alive in 1957, a tiny beeping object in the sky seemed like the ultimate in technological advancement—and quite possibly was a signal that the Cold War was taking on a new and terrifying space-based dimension.

We are very familiar with the people who brought us NASA's achievements, but there aren't many who would recognize the name of Sergei Pavlovich Korolev. He led the drive to produce the R-7 rocket, justified to the Soviets as being a platform for delivering nuclear missiles—this was the world's first true intercontinental ballistic missile—but in Korolev's eyes, the R-7 was always primarily a route to begin space exploration. Korolev was a man who the physicist Andrei Sakharov called driven and ruthless—yet that same drive meant that he ensured his teams were well-provided for at time when many were suffering hardships in the Soviet Union's harsh regime. To help with his vision, Korolev wanted the best from his workers.

Long term, Korolev's plan was to build large space stations in near-Earth orbit with the goal of gaining enough experience of working in space to repurpose the same structures as interplanetary

spaceships. This isn't as crazy as it sounds—the distinction between a space station and a deep-space ship, without the requirements to cope with atmospheric resistance and the pull of gravity
is small. The only real distinction is the size of the engines and fuel
tanks. Space stations usually have small engines for modifying an
orbit, but the ship would need larger engines and fuel supplies to
enable the kind of burn that would achieve sufficient speeds to
cross the solar system. But initially Korolev was not looking so far
ahead and had only one focus. He intended to use an R-7 rocket to
get an artificial satellite into space.

On October 4, 1957, thanks to Korolev, the world changed.
Sputnik 1 was just 58.5 centimeters (under 2 feet) across, a fragile
metallic ball sprouting four antennae that was humanity's first true
venture into space, propelled by one of the new R-7s. *Sputnik*'s
83-kilogram (183 pounds) mass, 61 percent of which was its batteries, made waves totally out of proportion to its negligible capabilities. Primarily political in the race between the United States and
the USSR, with a tiny scientific role, Sputnik made what had, until
then, been pure fiction seem real. This change in attitude is easy to
underestimate with hindsight. Back in the 1950s there were well-
respected scientists and engineers who considered it as unlikely
that we would get a satellite into space as it was that we would perfect time travel or build matter transmitters. They believed the idea
to be pure science fiction.

Of course, getting a small beeping metal ball into a three-
month orbit (the transmitter gave out after just twenty-two days
when its batteries died) wasn't exactly manned spaceflight. Nor
was it the kind of platform that could support orbital weapons, a
real fear for both sides at this point in the Cold War. But in less
than a month the USSR was at it again, turning a stray dog into a
world famous personality. On November 3, 1957, Laika was boosted
alive into orbit and into the public eye in a half-ton satellite. Sadly
the cute-looking, terrier-like mongrel died within a few hours as

Sputnik 2 overheated, but those who had claimed that it would be impossible to sustain life into orbit were proved wrong, and the race was on to get a human being up there.

CREATING NASA

First, though, the United States had to catch up with a simple satellite. Things did not start well. With all the pressure to match or better the triumph of *Sputnik 1* and *2*, all eyes were on the American Vanguard launch on December 6, 1957, which it was hoped would carry the first U.S. satellite into orbit. The public did not respond well when the rocket made it all of four feet into the air before crumpling and bursting into a fireball. It seemed as if there was no way to keep up with the Soviets. Learning the lessons of the previous December, when another attempt was made with Explorer 1 on top of a Jupiter C booster in February 1958, there was no live TV coverage allowed, just in case. But this time things went smoothly. The strangely shaped satellite (it looked like a miniature rocket) made it into orbit and the space race was on with a vengeance.

The initial work on satellite launchers, intimately tied as it was to ICBM design, came from the military, but the decision was made in the United States to split military and civilian scientific space development into two. The President's Science Advisory Committee recommended to President Dwight D. Eisenhower that a new civilian space agency should be set up, based on the foundation of NACA, the National Advisory Committee on Aeronautics, a small body dating all the way back to 1915. As a result, on July 29 1958, a bill was passed into law bringing into force the National Aeronautics and Space Administration—NASA.

However, the key year that shaped the future of manned spaceflight for decades to come was 1961. It was then that, despite all U.S. efforts, the USSR succeeded in being first again with Yuri Gagarin's

Vostok 1 mission. This made him the first human to make a true spaceflight (for a single orbit) on April 12, 1961. At less than two hours, his flight was shorter than some of the tourist hops out of the atmosphere planned by modern commercial space ventures like Virgin Galactic (see page 91), but it was still the first in history. America's Alan Shepard followed very soon after in *Freedom 7* on May 5, though his suborbital flight only lasted fifteen minutes and was arguably more a technology test than a true mission. (Shepard jokingly remarked, "They wanted to send a dog, but they decided that would be too cruel.") The first American to orbit the Earth was John Glenn on board *Friendship 7* on February 20, 1962. But by then, everything had been changed by a speech.

THE WORDS THAT OPENED UP SPACE

President John F. Kennedy probably had more "flashbulb moments,"—key shared memory events—in his short presidency than has any other incumbent of the White House. Yet for spaceflight enthusiasts there was only one such incident that changed the world. There is little evidence that Kennedy had any particular interest in space exploration before this. It hadn't been high on his agenda. But as a result of recent events, the time was politically right to have a program that would unite the country (and put the USSR in its place). On May 25, 1961, just twenty days after Shepard's fleeting flight, Kennedy made his Moon speech in the stadium of Rice University in Houston, Texas.

The most famous line that is often remembered from the speech, pledging a manned landing on the Moon in the 1960s, doesn't actually exist. The closest to it is this:

> However, I think we're going to do it [get a manned rocket to the Moon and back], and I think that we must pay what

needs to be paid. I don't think we ought to waste any
money, but I think we ought to do the job. And this will be
done in the decade of the sixties."

But there was much more in this stirring message. For a start,
there was a call to remember the pioneering spirit that made
America, and that is central to space exploration.

Reflecting on the pace of change, Kennedy commented:

This is a breathtaking pace, and such a pace cannot help
but create new ills as it dispels old, new ignorance, new
problems, new dangers. Surely the opening vistas of space
promise high costs and hardships, as well as high reward.
So it is not surprising that some would have us stay where
we are a little longer to rest, to wait. But this city of
Houston, this State of Texas, this country of the United
States was not built by those who waited and rested and
wished to look behind them. This country was conquered
by those who moved forward—and so will space.

Kennedy assured those present that the exploration of space would
go ahead, providing one of the greatest adventures of all time, "and
no nation which expects to be the leader of other nations can expect to stay behind in the race for space." That was a key line in the
speech, underlining the *political* significance of any space achievements. This was something that would still have been raw after
Sputnik and Gagarin. America was behind in the race for space,
but it couldn't afford to stay there. Kennedy emphasized just how
much the Cold War was extending from the surface of the planet
out into space.

[S]pace science, like nuclear science and all technology,
has no conscience of its own. Whether it will become a

force for good or ill depends on man, and only if the
United States occupies a position of pre-eminence can we
help decide whether this new ocean will be a sea of peace
or a new terrifying theater of war . . . There is no strife, no
prejudice, no national conflict in outer space as yet. Its
hazards are hostile to us all. Its conquest deserves the best
of all mankind, and its opportunity for peaceful cooperation
may never come again.

THE EAGLE REALLY DID LAND

It is hardly necessary to retell the details of the Apollo program
itself, leading to the successful landing of *Apollo 11*'s "Eagle" module in the Sea of Tranquility on the Moon on July 20, 1969. Neil
Armstrong and Buzz Aldrin became the first humans ever to arrive on the surface of another body in the solar system. There, with
great symbolism for this final and triumphant reversal of the space
battle with the USSR, the U.S. flag was planted for the first time on
another world, not to claim it—the Outer Space Treaty prevented
that—but to mark an American triumph.

A technical error with that first flag on the Moon was responsible for one of the many bizarre conspiracy theory arguments
that human beings have never undertaken the journey. Some believe that because the United States had to get there first and this
was proving technically difficult, rather than actually fly to the
Moon, NASA made a mock-up of the landing in a studio. Enthusiasts for this theory look for inconsistencies in what we see in photographs and videos. Because there is no air and hence no wind on
the Moon, you would expect the flag to hang limply, yet in fact it
flies boldly. "Evidence of fakery," mutter the conspiracy theorists.
In reality, it was just a matter of a thin rod extending along the top

edge of the material (and clearly visible on high-resolution photographs) to hold the flag up. The technical error was simply that the rod wasn't long enough for the width of the flag, so purely by accident the material had a big ripple in it, making it look as if it was indeed blowing in the wind. The error was intentionally repeated on later missions as the accidental effect looked so good.

Quite why there are so many conspiracy theories about the Moon landings, as opposed to any other well-recorded major event in history is not clear. After all, it's not as if getting to the Moon is rocket science. Okay, it is rocket science, but it isn't very *complicated* science. The basic calculations to set up the flight used mathematics that dated back to Isaac Newton in the seventeenth century. The rocket technology was simply a relatively small advance on those wartime A-4 rockets behind the V-2 weapons. And this lack of complication extended to the technology on board the Apollo missions. This was so unsophisticated, it seems staggering today. The onboard computer, for instance, had the equivalent of 4,096 bytes of RAM as its only writable memory. That's between 4 and 16 million times less than a typical smartphone today.

The basis for the conspiracy theories seems to be because the United States needed to win the space race with the USSR after being beaten on the earlier firsts—and hence to fulfill Kennedy's prophesy, made all the more essential after the president's assassination. Since NASA wasn't up to it, claim the conspiracy enthusiasts, the authorities faked it instead. But there is no evidence that NASA wasn't up to it—and to fake the Moon landings and not have good evidence of this leaking out would seem harder than doing the real thing. Imagine the vast numbers of people who would have to have been in on the secret, but who never admitted it. Most of the supposed evidence for fakery, like the "fluttering flag" is down to misunderstandings of the science, or basic errors, like claiming Neil Armstrong couldn't have been first on the Moon as someone photographed him stepping onto the surface. It seems crazy that anyone would think that NASA could make a Moon lander, but couldn't attach an automatic camera to it.

WHO WERE CONRAD AND BEAN?

It is interesting and very telling for the future exploration of space to see just how quickly public interest and memory dropped off after the original achievement. In part thanks to the movie, we all remember *Apollo 13* and its near disaster, and could probably name James Lovell, the astronaut played by Tom Hanks in the movie. But out of interest I took an online straw poll and found very few people who could tell me why Charles Conrad and Alan Bean were famous. Yet these were the third and fourth people to set foot on the Moon. It is in the nature of pioneering that "more of the same" doesn't play well with the crowd. It is arguable that there won't be names as well remembered as Armstrong and Aldrin until we get the first Moon base, or the first humans on Mars. We need the next great adventure to really register in the human psyche.

Once the series of Apollo missions had finished with *Apollo 17* in 1972, manned spaceflight had reached a peak that is yet to be surpassed. There was little point in continuing the missions because the goal was to get there first, not to make anything of a lasting human presence on the Moon, whether it was to do serious science or to establish a Moon base that could eventually become a colony. Kennedy had said that there would be a manned landing on the Moon in that decade, and there was. Mission accomplished. Unfortunately this also meant a drastic retrenching from true exploration to playing in our own backyard. Yes, we have had people in space for much longer periods on the ISS now, but since 1972, the baton of useful exploration and advancement has passed back to the unmanned missions that began with Sputnik.

THE IRON RING

Thousands of satellites now ring the Earth, and though many of them are defunct, others perform invaluable services day by day. It is they, rather than any manned missions, that have provided scientific and commercial reward for our ventures into space so far. Pointing toward Earth, we have satellites providing the GPS navigation system, collecting data for weather forecasts, mapping the surface of the planet and providing worldwide communications and TV broadcasts. Facing away, satellites have replaced the great observatories as the best way to make many of our astronomical observations. Few have failed to be inspired by the awesome images produced by the Hubble Space Telescope, while other satellites like COBE, WMAP, Herschel, and Planck have made use of a whole range of different energies of light and have provided unrivaled information on the state of the early universe in the form of the cosmic microwave background.

Unmanned probes have also scoured the solar system. In these ventures, the Moon and Mars have fared particularly well as destinations, but others have taken in Venus and headed in toward the Sun, or taken a new trail outward to give us our first detailed images of the outer planets and their remarkable moons. *Pioneer 10* and *11*, followed by *Voyager 1* and *2* have sent back unrivaled pictures of the outer solar system and *Voyager 1* is now the furthest traveled of any human artifact, reaching the edge of the solar system at around the time this book is published.

EXPLORING THE SOLAR SYSTEM

Attempts to explore beyond the Moon began as early as 1960 with a series of failed Russian Mars probes that mostly didn't make it out of Earth orbit. The earliest successful flypast of Mars was undertaken by NASA's 1964 *Mariner 4* mission, which gave us our first close-up view of the planet. Although notable as one of the first chances NASA had to get a lead on their Soviet counterparts, and at that the first mission to come close to another planet, the images sent back by *Mariner 4* proved a disappointment. We had hoped to find on Mars something close to Earth's surface, a planet that suggested the possibility of life as we knew it, but instead, the photographs beamed back were of a battered Martian landscape that was far more like the surface of the Moon than anywhere on Earth.

The first probe to survive landing on Mars was the Soviet lander *Mars 3* in 1971, which managed to communicate with Earth for all of twenty seconds before falling silent, leading to worried speculation that the Martians had sabotaged this invader. This was proved unlikely when in 1973 *Mars 5* lasted nine days on the surface, but the Russian attempts would soon be eclipsed by a whole host of NASA missions. These began with the Mariner probes, but moved up a gear with *Viking 1* and *2*, both launched in 1975 and both successfully sending landers down to the surface in 1976. More recently we've seen the likes of *Spirit, Opportunity,* and *Curiosity* not just land on Mars but rove around the surface, taking in a much wider area of the Martian landscape.

The USSR did better when it came to pioneering successful probes of our other neighboring planet Venus, which it has to be remembered was until the 1960s thought of as the most likely planet in the solar system after Earth to be able to sustain life. It clearly had a thick atmosphere, because all that is visible through telescopes is white cloud, and obviously it was going to be hot, because it is significantly closer to the Sun than Earth. But no one anticipated the hell that it actually was. A series of Venera probes from 1961 onward gathered data on flyby missions, while the first successful lander was *Venera 7,* which in 1970 became the first man-made probe to land in one piece on another planet, a year ahead of its Martian equivalent. It lasted all of twenty-three minutes in the harsh conditions of Venus, only managing to send back the surface temperature—but that in itself was enough to start the change of image of Venus away from a tropical paradise to the hellish heart of a metal foundry.

On the whole, landers for Venus had a better chance of surviving the descent to the surface because the thick atmosphere meant that parachutes proved very effective in slowing the descent of a probe. If a lander made it to the surface, though, it was much less likely to keep working for long on Venus than one would on Mars, as the combination of crushing atmospheric pressure and extremes of temperature mean that the survival capability of equipment is precarious at best. Most Venusian probes have given out within an hour of landing, sending data back frantically before they expire.

American probes arrived at Venus soon after the USSR's first efforts, with *Mariner 2* achieving a flyby in 1962 and further Mariner visits in 1967 and 1973, the latter, *Mariner 10*, being more significant for flying on to be the first probe to visit Mercury. More detail was revealed by the shuttle-launched Magellan mission in 1989, which produced radar maps of the surface of Venus and the European Venus Express in 2005, which is still active at the time of writing. Heading in the other direction, *Pioneer 10* and *11* were the first to visit Jupiter, and in *Pioneer 11*'s case, Saturn too before heading out toward the edges of the solar system. Launched in 1972 and 1973 respectively, *Pioneer 10* stayed in contact for thirty-one years, while *Pioneer 11* shut down twenty-two years into its flight.

The Pioneer craft were eclipsed, though, by *Voyager 1* and 2, both launched in 1977 to make use of a rare opportunity of planetary alignment that meant that they could take in not only Jupiter and Saturn but Uranus and Neptune too, while getting close-up studies of many of the larger moons that festoon the outer planets. *Voyager 1* concentrated on the inner two of the gas giants and at the time of writing is still operational as it heads toward the very edges of the solar system, making it our furthest traveled probe to date. *Voyager 2* was confusingly the first of the two to be launched but was overtaken by its speedier sibling as *Voyager 2* dallied with Uranus and Neptune after the inner gas giants. *Voyager 2* is also still active in 2013 (at the time of writing) and comes a close third to *Pioneer 10* and *Voyager 1* in the lists of our long-distance probes that will eventually leave the solar system. The Voyagers have been followed up by further probes—Galileo, Cassini, and Huygens—all adding considerably to our information about the outer planets and their satellites. Huygens, which traveled with Cassini, was probably most remarkable in making a soft landing on Saturn's moon Titan, where it survived for around ninety minutes.

HELLO ALIENS, EARTH CALLING

All the probes destined to exit the solar system have carried some kind of attempt to communicate the nature of humankind to any aliens that should come across these tiny beacons of human capability. *Pioneer 10* and *11* had a gold-plated plaque showing our location in the solar system, the appearance of a naked man and woman, and an attempt to locate our star using pulsars; while *Voyager 1* and *2*—reflecting a technology that has ironically all but disappeared from use on Earth—carried gold-plated metal discs, inscribed on one side with illustrations that are supposed to indicate how to play it, and the location of the Sun indicated by fourteen easily observed pulsars, and on the other with a groove like that on a vinyl record but including video images as well as audio. (The discs did not have the images of naked people after complaints from the public.)

If we are to believe well-known British physicist Stephen Hawking, these attempts to flag up our whereabouts could be a massive mistake on the part of humanity. Hawking has commented, "If aliens visit us, the outcome would be much as when Columbus landed in America, which didn't turn out well for the Native Americans," suggesting that any alien visitors could regard the Earth as simply a precious resource to mine, rather than a friendly civilization to visit. Commenting, "We only have to look at ourselves to see how intelligent life might develop into something we wouldn't want to meet," Hawking suggests we do everything we can to conceal our presence on the Earth from alien life, rather than draw attention to ourselves.

WHERE DID IT GO WRONG?

Taken in isolation, the achievements of the practical Earth-facing satellites and the probes that we have sent out into space are remarkable indeed. But set alongside the promise of human exploration that featured in Kennedy's speech it seems that something has stalled in our vision and reach. We should remember that the Moon was not the end point in Kennedy's vision. "The eyes of the world," he said, "now look into space, to the moon and to the planets beyond." Yet after the frantic scramble to be first to the Moon, followed by the apparently promising development of the space shuttle, which first flew in 1981, all inertia on manned flight was lost. (As we have seen, it is a curious coincidence that in James Blish's fictional future of his *Cities in Flight* series, dreamed up in 1956, it was also the 1980s when the space program stopped reaching further.)

Some would blame NASA for having goals that seemed driven primarily by the need to keep its workforce busy, justifying its budget, rather than making real steps forward. Others might point to financial restrictions, or the lack of true international cooperation that would make it so much easier to achieve really far-reaching goals. It's not that NASA has ever said, "We won't go back." Even in the hiatus that followed the *Challenger* disaster, NASA was planning for a permanent base on the Moon and a manned Mars program by 2007. The dreams have certainly been there, but the consistent political drive has not.

Space exploration is an expensive business that needs a vision that goes beyond the next election to take on the projects that are involved. Arguably one of the most important developments of twenty-first-century spaceflight is the increasing significance of commercial projects, where those backing the spaceflights are used to having longer-term plans than those of most politicians. Even so, it is unlikely that the commercial developments can be the sole driver for large-scale exploration, and ideally we should be looking for a fruitful coming together of the sort of large-scale international governmental investment that have made the ISS and, on Earth, the CERN particle physics facility possible with the speed, flexibility, and drive that can only come from the private sector.

An analysis of the typical life cycle of a failed NASA project clearly identifies the underlying practical problems in getting anything significant and new achieved. A mission is first proposed with dramatic goals—a manned mission to Mars, for instance. A mission that inspires the public and is likely to, at least initially, get political support. Along with this comes a proposed budget that should just about make it possible, as long as nothing major changes. But as more detail is filled in and the project takes shape, two things happen in opposition. The estimated costs rise, but the budget is simultaneously reduced.

Quite why this happens is complex. Costs tend to rise both because initial plans oversimplify and overlook expensive challenges, some of which are inherently impossible to predict, and because there is a natural tendency to play down costs to get a project approved. Budgets are cut because space exploration does not exist in a vacuum—when the economic environment means that money is tight, generally a space project will be less appealing that an expenditure that is more directly connected to the everyday welfare of voters. The only way to go forward is to reduce the scope of the mission. A manned landing on Mars becomes a flypast or even an unmanned probe. As the project continues, costs rise even further, the scope is cut once more and often the whole thing gets cancelled—because we now have an expensive project that has none of the emotional appeal and pioneering achievement of the original grand vision. It has lost all its allure.

The physicist Freeman Dyson, best known for his contribution to quantum electrodynamics, the science of the interaction of light and matter, experienced the workings and machinations of NASA firsthand when he became involved in the Orion nuclear-powered spaceship project (see page 208). Dyson drew a contrast between what he saw as the "real NASA" and the "paper NASA." Real NASA was driven by politics and finance. It was cautious, more inclined to cancel than commission, and struggled to manage its complex portfolio, leading to failed projects. The paper NASA, by contrast, was free-ranging, full of great ideas and truly inspiring.

This picture of contrasting NASAs is still as true today as it was when Dyson made the observation in the 1960s. There has never been a time since Apollo when paper NASA hasn't been planning returns to the Moon, has not set its sights on trips to Mars, and has not pursued exotic technology like space elevators and even warp drives for starships. But the real NASA, the one set about by politics and budgetary restraint, struggles to get anything novel off the ground. It would be unfair to label NASA worthless, but it never seems to bring the paper and real sides of its existence close enough together.

It is a sobering reality that the *majority* of significant space programs, particularly of planned manned missions since the 1970s, have been cancelled without coming close to fruition. As for that lack of emotional appeal that is felt for the watered-down missions that do take place, you only have to contrast the worldwide public response to the Apollo program, following its every development, hardly sleeping through the first Moon landing (and for that matter, keeping up with every nuance of the *Apollo 13* rescue) with the total public indifference to ISS missions.

Of course there are brief exceptions. The public is happy to get behind the ISS when, for instance, the station commander proves himself to be an entertaining singer of a David Bowie song. But this is a rare exception. The problem is not just a matter of familiarity and repetition. However original and useful the ISS missions may be (and there is considerable question about that), it is hard to develop any sense of pioneering interest about what is, in the end, a voyage upward toward space that is no further than the distance between Boston and Philadelphia. It's not exactly boldly going very far.

LIVING TO CHECK THE BOXES

This lack of impetus also helps contribute to the stultification of future missions. In an organization like NASA, when staff have had a number of missions cancelled without ever getting anywhere they can easily begin to operate in a "time server" mode, where their aim is not to achieve any breakthroughs in space, but rather to keep the whole paper-pushing organization going. The goal becomes keeping a job, rather than expanding the frontier. Such bureaucracy in space is parodied brilliantly in the science fiction short story "Allamagoosa," by the now largely forgotten but brilliant English author Eric Frank Russell.

In the story, a military space cruiser is due for a routine inspection and so the crew spend days checking off every item on the ship against the official manifest. To their horror, they can't find one of the items demanded by the military equivalent of head office. According to the manifest, they are supposed to have an offog on board. But no one has the faintest idea what this is. To ensure that they pass their inspection, the ship's engineers mock-up a functionless electronic device, which is presented to the visiting admiral as their offog. So far, so good—but the crew realize that this deception will not stand up to a detailed inspection by a technical expert back at port. So they report that their offog has unfortunately been broken during maneuvers, when it came apart under gravitational stress.

To the crew's surprise, all the ships of the fleet are immediately recalled for technical checks. It is finally revealed that the offog was a misprint for "off. dog"—official ship's dog. The Navy hierarchy were horrified to hear that the ship's dog had come apart under gravitational stress. Hilarious though the story is, this urge to ensure that the paperwork is correct, rather than take logical and useful action is a microcosm of the way that an organization can become obsessed with checking the boxes, rather than fulfilling its actual purpose. Repeatedly cancelled missions, never achieving the supposed purpose of NASA can only have contributed to this kind of inertial damping that must at least in part be responsible for the lack of progress since the 1970s. Again it is to be hoped that the smaller, keener commercial enterprises now planning to take on space travel can cut through these problems and perhaps bring NASA and its equivalents around the world up to the mark to enable a breakthrough in space exploration.

I ought to stress this isn't anti-NASA diatribe. The organization often does a brilliant job, and the same problems also apply to, for instance, the European Space Agency, which has the added problem of being at the beck and call of multiple countries, all attempting to direct its projects. The environment is changing, and the increasing involvement of private companies in not only constructing spacecraft but launching them should make a positive change. But it has to be expected that national and international agencies will continue to have a major role and to do so effectively there needs to be change.

If we compare the incredible successes of NASA in putting men on the Moon less than a decade from Kennedy's speech with later projects that have sometimes taken longer than this to achieve absolutely nothing, there is a clear lesson to be learned. We see budgets cut and goals modified until the whole project lacks any appeal and is cancelled. The only way to succeed with this kind of venture is to set a clear vision and stick to it, making commitments that you really intend to follow through with (which is a difficult concept for most politicians). This doesn't mean that rocket scientists and engineers should get free rein to spend whatever they like. Even when money is freely available it is important that politicians ensure that budgetary concerns are not ignored, and doubly so at a time of financial difficulties. But there has to be a long-term vision that is not constantly under attack if there is to be major progress in space.

GOING INTERNATIONAL

Another lesson can be drawn from the accelerator wars in particle physics. Particle accelerators are very expensive pieces of equipment. They typically consist of huge tunnels, miles long, with vast superconducting magnets and detectors, designed to push particles near the speed of light, then observe what happens when they crash into each other. This kind of facility does not come cheap. The United States planned to build a world-beater, the Superconducting Super Collider (SSC) in Texas to be first in the hunt for the Higgs boson. Like a failed space project, this was constantly under financial attack as estimated costs rose until it was finally cancelled after $2 billion had already been spent. Achievement: nothing. By contrast, CERN (Conseil Européen pour la Recherche Nucléaire) in Europe went on to build the Large Hadron Collider (LHC) with international support. It wasn't as good as the SSC would have been—but it has done the job and been a huge success.

We do indeed need to help and encourage private enterprise to be involved in space exploration, but some of the big projects will benefit from government involvement, and the SSC versus LHC contrast shows that one of the most effective ways to do this with these vast, expensive projects is with good international cooperation to share the load. One of the huge benefits here is that when one country has to reduce its commitment and can't pay as much (or pulls out altogether) one or more of the others can step up and keep the project going. This approach is much more robust in the face of economic and political difficulties—and provides a bigger pool of talent to draw on.

The downside of such collaborations is that they appear to reduce the patriotic content that is part of the political driver for space exploration and that was central to President Kennedy's speech. This concern is clear when you look at an interview between astronomer and space travel enthusiast Neil deGrasse Tyson, and TV host Stephen Colbert. In the interview, Colbert comments: "If we land on Mars, how are we going to know if U.S.A. is number one if an American astronaut is standing next to a French guy? Are we going to say, 'Go, Earth!'? No, we're going to say, 'Go, U.S.A.!' Right?"

It is arguable that what is needed here is a maturing of the idea of patriotism. Colbert's is the patriotism of the child. It is all about who is best. About identifying the "number one." What he fails to see is that it is perfectly possible to still say "Go U.S.A!" for an American astronaut, even if he is standing next to a French guy. (Or better still, an English guy.) Given Colbert's track record this is probably intended to be satirical, but it reflects a common enough sentiment. The fact that someone else has also succeeded doesn't mean that you have failed. In the grown-up world, win-win is better than win-lose.

With an international collaboration, it is possible to achieve much more and to do so with less strain on the budget. Yet this wouldn't prevent the United States from taking the lead in many areas, and the United States would inevitably be widely represented in terms of astronauts and other key figures. What is there to be scared of? International collaborations bring their own problems—there needs to be extra care to ensure that bureaucracy does not take over—but it is hard to imagine that they won't be the best way forward in the long term for the governmental side of space exploration.

JOYRIDE TO THE STARS

There is much talk these days of space tourism. This is not a totally new concept. Since Dennis Tito, the American engineer and entrepreneur behind Wilshire Associates paid $20 million for eight days on the ISS in 2001, a number of wealthy individuals have shelled out millions to join the regular space program. At the time this first happened, it was not popular with NASA. Daniel Goldin, then the administrator of the organization was quoted as saying, "We don't have time to hand-hold tourists that don't have the proper training." But the next generation of space tourists will be very different from traditional astronauts—and in many ways much less pioneers; more high-risk day-trippers.

Perhaps the best example of the developing (if mostly speculative) market in commercial space tourism—and certainly the best advanced in practical terms—is Virgin Galactic. While in all probability this was started by Richard Branson as one of the publicity stunts that Virgin is famous for, it has become a realistic commercial enterprise. At the time of writing over 530 would-be astronauts had signed up to join Virgin on a venture into space. With prices starting at $200,000 and rising to $250,000 for a seat this is still no mass-market venture, but it is being undertaken as a serious business. If it were a NASA project, this has now migrated from paper NASA to real NASA.

Virgin has based its venture on *SpaceShipOne*, the world's first private manned spaceship, which made it into space in 2004. The approach used is to carry an enhanced version, *SpaceShipTwo* up as high as possible using an aircraft, *WhiteKnight2* (I don't know why the aversion to spaces in these names). The rocket engine then takes over, powering *SpaceShipTwo* up for around 70 seconds to take it to a height of around 110 kilometers (68 miles) before gliding back down to Spaceport America in New Mexico. The whole experience is planned to last around 2.5 hours (so costing an impressive $1,666 per minute). This is a suborbital flight, but just about makes it into true space. By comparison, the ISS in a low Earth orbit is around 410 kilometers (250 miles) up.

In devising this service, Virgin has benefitted from the boost given to commercial spaceflight by the Ansari X Prize. This was a $10 million prize, offered for the first reusable manned space vehicle that could enter space twice within two weeks. An essential requirement was that participants were nongovernmental, probably one of the key drivers in the move away from space travel being purely the remit of NASA and its international equivalents. The prize was inspired by the early days of aviation which saw the Orteig Prize for the first aviator to fly nonstop from New York to Paris (won by Charles Lindbergh) and the Schneider Trophy, which resulted in the construction by British manufacturer Supermarine of the direct ancestor of the superb World War II fighter plane, the *Spitfire*.

Technically the Coupe d'Aviation Maritime Jacques Schneider, the Schneider Trophy was for a race with prize money between seaplanes and was intended to inspire the designers to push the envelope in aircraft design—which it certainly did. The Ansari X Prize took a similar approach for the vehicles necessary for easy suborbital spaceflights. The winner of the prize was Burt Rutan's design *SpaceShipOne*, built by Mojave Aerospace Ventures, a combination of Rutan's company Scaled Composites and financial backing from Microsoft founder Paul Allen.

The prize money was certainly an incentive—it was considerably more generous than the $25,000 offered for crossing the Atlantic (worth either $330,000 or $1.6 million now, depending on whether you consider simple monetary inflation or the purchasing power of wages). But the costs involved would have far exceeded this and the real driver was both the simple achievement and the likelihood that whoever succeeded would make money in upcoming lucrative space ventures, like Virgin Galactic—winning seems to have paid off for Scaled Composites.

Virgin isn't the only player in the space tourism market offering futures on trips into our space backyard. XCOR Aerospace comes in with the cheapest of the options for those hoping for a spaceflight on a budget, offering the single passenger seat on their *Lynx* suborbital flights for a mere $95,000. Unlike Virgin's two-stage process, *Lynx* is a single, rocket-powered craft that takes off using its own engines, and carries its passenger for a round-trip flight to the edge of space in around fifty minutes. At the time of writing, *Lynx*'s engine had been tested on the ground, but there had been no test flights.

Another big name in the market is Space Adventures, which despite sounding like a theme park ride has already been responsible for getting the likes of Dennis Tito and Charles Simonyi into orbit, an exclusive club soon to be joined by singer Sarah Brightman (who announced in 2012 that she intended to make the trip to the ISS). But Space Adventures also has an equivalent of the Virgin offering, to be operated by Armadillo Aerospace. This venture has a way to go before catching up with the already partially tested Virgin Galactic. More remarkably, Space Adventures is also already advertising the ultimate in wish-fulfillment space tourism, a place on a hypothetical lunar mission—not quite as exciting as it sounds as there would be no landing, just a trip around the Moon. This puts the dreaming (if not the practicalities yet) into a whole different league to the suborbital flights—and is likely to involve costs that make the price tag for Dennis Tito's trip seem like small change.

These are the most advanced of the players but a whole range of other companies have opened with the intention of providing individuals the chance of getting into space. Some are, as yet, purely paper companies—others have already flown prototypes. At this stage, though, beyond Virgin and Space Adventures it is difficult to be sure who the winners will be. There have also already been a number of failed attempts where companies have made a start but fallen by the wayside. These are not just overenthusiastic amateurs but include companies like Beal Aerospace and Rotary Rocket, which both got as far as testing prototypes, losing millions of dollars—and there are bound to be more failures to come.

This change in approach to space travel will also mean a different attitude is necessary toward would-be astronauts. We are used to our space pioneers being human specimens at the peak of physical ability, often chosen more for their military expertise as pilots than for their insights as leading scientists. Clearly such a selective approach based on physical stamina is not going to work for space tourism. While there is likely to be some form of medical cutoff, particularly checking the customers' ability to cope with high g, which can have an impact for those with blood pressure or heart problems, there is no expectation that would-be sub-astronauts will be hugely fit, or that they will be restricted by height and weight limitations as has happened with the real thing.

Similarly, the amateur astronauts can expect to undergo a degree of training, but nothing like the intensive and selective program that a traditional astronaut is put through. It is unlikely that any commercial passengers will fail their preflight training. What they experience will be more akin to the kind of routine that skydivers undertake before taking their first jump. Virgin offers the most extensive training with three full days building up to the experience, including evacuation drills and high g training, while other companies recommend that passengers do some of their own preparatory work by taking zero gravity flights to experience weightlessness.

THRILLS AND SPILLS

There now seems little doubt that commercial spaceflights will carry space tourists in the near future. Leaving aside the handful who have already flown, who have just bought in to the official ISS program, services like Virgin Galactic will certainly begin to take passengers out to the edge of space. This is a strange hybrid of the frontier and the theme park. Participants will have none of the pioneering drive, none of the real advantages of being on the frontier, yet they will face some of the risks that are inherent in traveling at the edge of our technical ability.

Any attempt to make it impossible for a trip into space to fail encounters the inevitable problem that there are so many things to go wrong—and sometimes a seemingly minor technical problem can be catastrophic. It is to be hoped that the designers of all the commercial spaceflights, from Virgin Galactic to the planned commercial Mars missions (see page 148) have studied the disaster that befell the Russian craft *Soyuz 11*, the first to visit the then new Salyut space station, as it made its descent to Earth on June 29, 1971. The nightmare began with a hissing sound as the descent capsule detached.

The three-man crew unbuckled their seatbelts and began searching frantically for what they believed to be a leak. It was difficult to pinpoint the source of the sound with their restrictive headgear, but eventually they discovered that a valve under one of the cosmonauts' couches, designed to equalize pressure inside the cabin with the outside before opening the main hatch after landing, had accidentally been opened while still in flight. A tiny opening, about the size of the quarter, the valve was draining their air into space. The cosmonauts tried to close the valve manually, but could not manage to do so. Within two minutes Georgi Dobrovolsyi, Vladislav Volkov, and Viktor Patsayev were all dead. One small leak is all it takes for a fatal accident when you venture into space.

As the space shuttle disasters have also demonstrated, attempting to leave the Earth stretches our technology to the limits. These are very complex operations, with a lot that can go wrong. Just a small problem can have catastrophic effects. Estimates of the risk of dying on shuttle missions were typically between 1 in 20 and 1 in 500 (the latter generally thought to be far too optimistic). This is clearly not a level of risk that is acceptable for a commercial tourist flight. And although the technology used in suborbital flights is far simpler than it is in a shuttle, these craft have to cope with conditions far beyond those of basic aviation. Even the technology itself is more dangerous. This was demonstrated all too painfully in 2007, when an explosion during a test of *SpaceShipTwo*'s engines resulted in three deaths and more injuries.

There is no doubt that commercial operators will have to go through rigorous safety checks, at a level far exceeding that imposed on NASA, where it was deemed acceptable to take greater risks for a national program of exploration than would be taken for entertainment. However there has to be a slight question mark over the mass marketing of a venture that is as inherently risky as voyaging to the edge of space. When the first serious incident occurs, it could change the attitude to space tourism forever. This is not a matter of being pessimistic. As executive director of the Commercial Spaceflight Federation based in Washington, D.C., Alex Saltman commented bluntly, "We know at some point an accident will occur."

When individuals choose to open a new frontier there are inevitably risks—but those risks are seen as being acceptable when put alongside the potential rewards, whether those be political or commercial. But the half-and-half venture of space tourism is one that will find it difficult to justify those risks for a simple thrill. Time will tell whether risk aversion (and the associated fear of litigation) or the urge to have a go whatever the dangers will win out.

THE WHEEL IN SPACE

From the viewpoint of exploration by humans, the best we can do in the backyard of orbit is to put up a space station. As we've seen, historically and in science fiction there have been grand visions of great wheels in space, rotating to provide the occupants with gravity—where the reality has been more like hauling portable office trailers into space. Our space stations to date, whether the Russian Salyut and Mir stations, Skylab, or the current ISS have all been little more than assemblies of space capsules, like building a house from a pile of shipping containers. Perfectly possible, but hardly elegant or permanent.

There is a good reason why the real-life space stations bear little resemblance to the magnificent rotating wheels of fiction, first envisaged in detail in the 1950s by Arthur C. Clarke and German-American space visionary Willy Ley. It is simply down to the shape of the rockets used to carry materials into space. We continue to use rockets that are long, thin, and cylindrical. So space stations have been built of one or more cylindrical modules, because that's what fits in a rocket.

When the great wheels were envisaged, the idea was that a whole series of rockets would carry up the basic building material into space where space-walking astronauts would employ their special inertia-free riveting guns and wrenches to construct the massive structure, like construction workers on some space-based office block. But this approach was blocked both by the relative unavailability of regular shuttle traffic to carry up those materials and the difficulties of putting in man-years of work in space on the assembly, requiring a good sized space-based construction crew for many months. It was so much easier to send up one or more modules, if necessary join them together in a quick exercise and end up with one of the messy constructions that have become familiar to us.

It didn't help that the maximum capacity of the lifting rockets available to NASA was reduced after the 1970s. Skylab, the first American space station, was hauled aloft by the last of the Saturn V rockets that had been used for the Apollo program and which designer Wernher von Braun had intended to be the workhorses that would take human beings to the planets. Skylab's launch on May 14, 1973 marked the end of the program. And though the power of that launcher meant that Skylab could be four times the size of its Soviet competitors, the Salyut stations, a single launch was not enough. It is imaginable that a series of Saturn Vs—or modern equivalents—could haul up a series of slightly curved sections to be assembled into a vast wheel, but since 1973 we have not seen a launcher with the same capacity as Saturn V, let alone a bigger and better alternative.

The space stations we have so far constructed are useful as laboratories to experiment on the endurance of humans in space—to see, for instance, the impact of a long-term stay in space on human physiology. The record is currently held by cosmonaut Valeri Polyakov who spent 438 days on Mir, which seemed to have no long-term impact on his health. Polyakov undertook his extended flight in 1994—at the time of writing, nineteen years later, Polyakov is seventy-one and still going strong. Such piecemeal local stations could also provide crude space hotels for the commercial trade. But apart from a potential role as stopovers en route to a serious destination, the current type of space station does not provide any movement of the frontier. They are tied to Earth and to our spatial backyard.

The same was true throughout all the ups and downs of the Space Shuttle program. Although the theory was that this was a vessel that would slash the cost of space travel and see missions launching with the frequency of a commercial flight, the reality was that the shuttle, and NASA's instance on using it as America's only launch platform in an attempt to reduce overheads, was a drag on the development of manned spaceflight. Impressive though it looked, the shuttle was never more than a backyard toy. Quite different vessels are likely to be used to take on the wider solar system, or to explore further still. And, of course, the shuttle has now been retired.

THE TAU MISSION

To date, most of our automated missions we have sent out into the solar system have been fairly small in scope. However, just as NASA often had more grandiose ideas for manned missions that never came to fruition, so there have been paper-based projects that were intended to do much more—most dramatically in the TAU mission.

As far back as 1976, the Jet Propulsion Laboratory proposed an operation departing in 2000 that would burst the limits of the backyard, spending up to fifty years exploring beyond the edge of the solar system in the 50 to 150 billion kilometer from the Sun range (30 to 90 billion miles). By comparison, the outer region of the solar system known as the scattered disc reaches out to around 12 billion kilometers (7.5 billion miles) from the Sun. But some of the more extreme orbits of the bodies in that zone take in the range of this TAU mission (named for "Thousand Astronomical Units," where an "astronomical unit" is the average distance from the Earth to the Sun). These include Sedna, a minor planet which comes close enough to enter the solar system but also travels out to the 150 billion kilometer range.

To put the scale involved in context, at the time of writing *Voyager 1* had reached about a quarter of the distance to this region, so Tau was a fair challenge for 1976. The aim of the mission was not only to explore, but to test out long-range propulsion technology. However, it has become clear, with nothing more than a plan over a decade after the intended launch date that this is not going to be an easy mission to get off the drawing board. It may be, given the cost and timescale involved in any such major mission that it is more politically appropriate to aim for the stars and start off on an interstellar mission—even if it never reaches its destination—because it inevitably is going to engage the public's imagination far better than a mission that spends fifty years going nowhere, merely wandering just outside the solar system.

Even back in 1976 it was clear that a traditional rocket motor would not be appropriate—especially if Tau was to be able to access power for fifty years. The intention was to use a nuclear electric device, where a nuclear power source is used to generate ions and produce the electric field required to power an ion thruster. It was suggested that the ions themselves might be of mercury, though it might have been better if the reaction material could have been harvested in space in some way to avoid having to carry all of it on the probe.

THE UNWANTED VISITORS

The TAU mission apart, we think of sending probes to a specific location—to reach the Moon, Mars, or a known asteroid. Even this picture is a little fuzzy as none of these bodies have fixed positions with respect to the Earth. We are aiming for a moving target, which is particularly obvious when aiming for high-speed visitors like comets. However, there is another class of solar system body that is less friendly than the typical comet and that we don't have to head out too far to visit. These are the near-Earth objects. As the name suggests, near-Earth objects are asteroids and comets that come into close proximity and that have the potential to get closer than is comfortable. The Earth is constantly being bombarded with tiny bits of rock and metal that burn up in the atmosphere. Around 100 tons of material blast toward us and burn up every day. But occasionally something larger comes in our direction and the outcome can be catastrophic.

Perhaps the best-known disastrous encounter we've had with a near-Earth object was the one that is thought to have been a major contributor to the extinction of the dinosaurs around 65 million years ago. An asteroid around 10 kilometers (6 miles) across appears to have hit the Earth, causing massive damage (probably impacting near Chicxulub on the Mexican Yucatán peninsula where there is a huge crater from around the right date) and most significantly for the dinosaurs, altering global climate for several years with the ash and debris that was thrown up into the atmosphere.

It might seem little more than guesswork to suggest the cause of something that happened so long ago with little direct evidence, but physicist Luis Alvarez and his geologist son Walter published a paper in 1980 that made this claim based on geological findings. In many parts of the world, along the boundary in the Earth's strata that dates back to this period, there are unusually high levels of the metal iridium. There isn't much of this on the Earth's surface as it is heavy enough to have sunk down toward the Earth's core during the planet's 4.5 billion-year lifetime. But in asteroids the metal is not hidden away and so it was blasted across the Earth's surface on impact.

To give an idea of the impact of such a massive asteroid, it would release comparable energy to a nuclear bomb the size of the Hiroshima explosion going off ten times a second for around twelve years. Such huge collisions are rare, but significant enough to make it worth watching the skies for incoming near-Earth objects and to plan for mechanisms to divert them from collision with the Earth. To keep the scan to a manageable scale, the focus is on "potentially hazardous objects"—those that are 30 meters (100 feet) across or larger (so liable to cause widespread destruction) and that come within 7.5 million kilometers (4.6 million miles), which is close enough that they could be deviated sufficiently to hit us if they come too close to another planet.

After a couple of decades of programs attempting to pin down these potential visitors, we now know of thousands of near-Earth objects, of which perhaps 10 percent are large enough to be dangerous. Luckily we would only expect an object on a scale of a kilometer or more, capable of wiping out as much as 20 percent of the human population, to impact once every million years—but of course, this doesn't mean that it is impossible for this to happen in our lifetimes. It is highly unlikely, but it is possible. So far none of the objects discovered on this scale is predicted to present a threat in the next hundred years.

There have been several probes already that have visited near-Earth objects. NEAR (Near Earth Asteroid Rendezvous) Shoemaker was launched in 1996 and studied the large near-Earth object Eros, the first of its kind to be discovered, back in 1898. The NEAR probe orbited Eros and finally landed on it in 2001. Another notable first was the Japanese probe *Hayabusa,* which launched in 2003, landed on the rocky near-Earth asteroid Itokawa in 2005 and returned to Earth with samples in 2010. But these missions have just been a matter of observation. A realistic possibility is that, should we detect a seriously dangerous near-Earth object on the wrong course, it would need to be intercepted.

We already know of one near miss to come. On June 19, 2004, the Asteroid Survey telescope on Kitt Peak, Arizona, spotted an asteroid that would later be named Apophis. The risk of a near miss slipped attention until December, when alarm bells rang as it seemed that there was a 1 in 37 chance that this 300-meter (1,000-foot) asteroid, weighing millions of tons was on a collision course with Earth. As the data was refined this risk subsided, but it is still expected that Apophis will come within 30,000 kilometers (18,600 miles) of Earth. This will make it closer than some of our satellites—geostationary satellites that are in a fixed position over the Earth, for example, are around 15 percent further from the surface.

AVOIDING DESTINY

The potential for a disastrous impact means that a lot of thought has been put into ways of preventing an asteroid colliding with the Earth, should we be made aware of the danger in time. Changing the path of a potential impactor is not something that could be done overnight. The inertia of a large asteroid is such that it either requires intervention when it is many months away, so a small change in direction can produce a large change in final position, or it needs a huge amount of energy to be applied to produce a big enough shift.

The simplest approach is just to hit the asteroid with something. We have a small amount of experience of this—at least enough to be sure that the approach is navigationally feasible—when in 2005 the *Deep Impact* vessel sent a probe intentionally crashing into comet Tempel 1. This was too small a device to have any noticeable effect on the comet's trajectory, but if, for instance, you hit a 200-meter (660-foot) asteroid with a 5-ton probe traveling at an achievable 10 kilometers per second (22,300 miles per hour), it would be deflected by twice the Earth's radius in ten years. Catch it far enough out and such an asteroid could relatively easily be pushed off course. (This assumes, of course, that we are confident of its path for the next ten years, or our efforts could deflect it onto a collision course, which would be worse than embarrassing.)

Enthusiasts for nuclear bombs have long felt that an explosion, particularly a fusion blast, would be an ideal way to deflect an asteroid, whether the bomb was placed a little distance away and produced a pulsed nuclear motor style push (see page 206) or whether it was embedded in the asteroid to blast it into fragments that head away from the Earth. This approach may well work, though opportunities to test the technology are limited and the implications of getting it wrong are scary. What certainly isn't a good move is Hollywood's favorite approach of blasting the incoming asteroid into fragments just before it enters the Earth's atmosphere. This may make for a good cliffhanger in a story, and gives an opportunity for spectacular special effects, but unfortunately all that is likely to happen is that the pieces will still be too large to burn up and will produce a similar level of destruction to the original whole, but spread over a wider area.

Either way, the hope is that there will not be a large asteroid on a collision course with Earth until our technology is significantly more advanced. However, even though there are holes in the detection system and most of the mechanisms for deflection are untried and carry some level of risk that they won't work, at least we are a step forward now on all the previous millennia of human existence, when a large-scale impact would have produced an unbeatable endgame.

Whether we are venturing out to the ISS, sending unmanned probes around the solar system, or trying to ward off stray asteroids, we remain very much with an Earth-centered viewpoint. Just as in the Renaissance it became necessary to move away from an Earth-centered view of the universe to a bigger picture, so we need space travel to move on if we are to achieve anything greater. Perhaps the biggest shift will be when we move from getting people to the Moon or another planet as a temporary expedition before returning home to Earth, to establishing a new place for people to live permanently. To truly expand the frontier we need to set up a colony in space.

6.
FRONTIER COLONIES
IIIIIIIIIIIIIIIIIIIIIIIIIIIIIIIIIIIIIIIIIIIIIIIIIIIIIIIIIIII

If the desert remains—in popular imagery and, at least, the last earthly frontier, then by the same reasoning, outer space is the last unfulfilled locus of human destiny. Space "colonies," a term that is unfortunate, recalling the imperialism of earlier decades—are the ultimate step in the progression of visionary communities and superstructures.

—*Yesterday's Tomorrows: Past Visions of the American Future* (1996)
Joseph J. Corn

When the early explorers sailed to the shores of new continents they could only briefly visit the coasts before making the journey back home. This is very much our experience with manned space travel to date. Our astronauts spend a few days away from the Earth and then return. Even the ISS is little more than a way of camping out in the Earth's backyard. To really take on the frontier we need to establish colonies beyond the Earth's gravitational well. It is only then that we can truly widen our horizons.

LOSING THE URGE

In the early twenty-first century it seems that we have lost the impetus required to take on the missions necessary to lay the groundwork for any space colonies. As we have seen, the journey to the Moon in the 1960s took around eight years from conception to the *Apollo 11* landing. Yet in the far longer period since the Apollo program ended, we have failed to take our astronauts beyond a tight orbit around the Earth. At the time of writing, the advanced nations are almost all in recession and it is hard to envisage any government with the possible exception of the Chinese undertaking the spending required to return to the Moon or to take the next step and journey to Mars—yet for much of the period since 1969 finances have been relatively good. It wasn't a lack of funding that prevented manned exploration then, it was a lack of will.

Without doubt military and political drivers played a major role in making the Moon landings possible—but so did an outward-looking, enthusiastic approach from the very top. The Apollo program had a clear goal, explicitly stated by a charismatic president of the United States in the form of John F. Kennedy. Although there have been similar statements since, most notably President George W. Bush's 2004 speech (echoing a speech made by his father in 1989) proposing a return to the Moon and a manned mission to Mars, these fine words have not been backed up by either finance or by being made the fundamental goal of an agency like NASA. As we have seen, ever since Apollo, NASA has seemed to have a primary goal of keeping itself afloat, rather than aiming for the stars (or at least our neighboring planet).

LEGISLATING FOR SPACE

Having a long-term goal of setting up a colony was not helped by the Outer Space Treaty. Signed in 1967 by the United States, the USSR, and the UK it has since been joined by over a hundred countries and its aim is to control the activities of states in "the exploration and use of outer space, including the Moon and other celestial bodies." Without doubt, particularly in the climate of the Cold War when the treaty was written, this was a parcel of good intentions that bans the placement and use of nuclear weapons anywhere in space, and that aimed to make the Moon and other space bodies freely accessible to any nation. But the devil is in the detail and it is arguable that the treaty's restrictions, making the Moon, planets, and asteroids "not subject to national appropriation by claim of sovereignty, by means of use or occupation, or by any other means" has not helped enthusiasm for the idea of space colonies.

Is it even possible to set up a colony, let alone to get it to make commercial sense by, for instance, mining scarce raw materials if states are so restricted by the law? Look at the detail of the treaty, and the result is a confused picture. The aim of the treaty's authors was that the use of the Moon and other celestial bodies should be to the benefit of all countries, irrespective of their economic and scientific development. As a result it doesn't ban activity out there—in fact it explicitly states that these resources should be "free for exploration and use by all States without discrimination of any kind," but at the same time it specifies that there should be free access to all areas by everyone.

So what this seems to say is that yes, you can set up a colony on Mars, or retrieve a scarce mineral from an asteroid. But what you cannot do is prevent anyone else wandering into your colony or waiting until you find that mineral and then setting up alongside you, chipping away at it. It isn't possible for a state (or presumably a company) to claim ownership of the Moon, an asteroid, or a region of Mars—but the treaty does not prevent building a base, colonizing a planet, or making uses of resources (with the requirement to avoid dangerous contamination). It merely makes the legal position interesting, to say the least, should it ever be put to the test.

Whatever the exact niceties of interpretation of the treaty it seems likely that in the desirable aim of quashing any imperialist inclination to add a new world as an extension of an existing state, it also reduced the impetus to take manned expeditions to new worlds, particularly while the driver for manned spaceflight was political posturing between two superpowers. Now that things have settled down a little, it is possible that this becomes less of a problem and more of an opportunity.

THE TRUE SPACEMEN

When thinking of setting up space colonies, it is natural to imagine the Moon, or Mars, or even the asteroids as appropriate locations, but one of the earlier scientists to give serious thought to the matter, physicist Gerard O'Neill, took a more radical view. He believed that no planet or moon's surface was good enough for the unlimited expansion that humanity deserved. Admittedly there was plenty of room out there to begin with, but if history has taught us anything it is that we are far too good at filling up the available space. Why go for the relatively few usable surfaces in the solar system on existing bodies, when there is so much more empty three-dimensional space in between?

The question that nagged at O'Neill and drove him far from his comfortable academic roots was this: Was it feasible to put colonies in space itself, on artificial structures? Do that and space opens up, with immensely greater opportunities in its vastness than would ever be available on existing bodies. In the key year of 1969, when the Moon first came within our grasp, O'Neill was a lecturer in physics at Princeton, and wanting to spice up his somewhat dull Physics 103 introductory class, put on a seminar to look at the question: "Is the surface of a planet the right place for an expanding technological civilization?" What began as an entertaining thought exercise for his students would become the driving force of the rest of his life.

There was no doubt that O'Neill's inspiration was in part taken from science fiction. He was an enthusiastic reader of the likes of Arthur C. Clarke, whose 1952 novel *Islands in the Sky* describes prototypical artificial space colonies. O'Neill's seminar was also perfectly timed to be boosted by all the printed material that had been inspired by the Apollo program, with plenty of future-gazing tomes presenting ideas of huge space environments for thousands of occupants, either as expansion room for the Earth or as a means of crossing interstellar space in the form of generation ships (see chapter 9). At the same time there was plenty of political and environmental concern about the condition of the Earth and our ability to continue with uncontained growth. This was the time when the environmental movement was first becoming a serious force. O'Neil's topic was a natural to emerge when it did.

What began as a teaching exercise soon blossomed to fill all of O'Neill's spare time—and then his professional life too. And unlike many of the future gazers of the time, he was not prepared to simply call for developments with hand-waving enthusiasm but little appreciation of the science and engineering implications. He began to make detailed plans for all that would be required to set up living environments in space. Not limited outposts like space stations, but permanent colonies, large scale and with a whole ecosystem on board so that they would not have to depend on being constantly supplied from Earth.

When we now look at this enormous task, the biggest problem that confronts us is the sheer expense and effort of getting all the material to build the environments—and the people and start-up resources to fill them—out into space. As we saw in chapter 4, getting out of the Earth's gravity well is not a trivial exercise. O'Neill started on his quest at a uniquely optimistic moment in the history of manned space travel. Apollo was a huge success and NASA was being extremely bullish about the future capabilities of the newly designed space shuttle. It would, they said, allow 32 tons of material to be carried aloft on each journey—and the shuttle fleet could make between thirty and fifty flights a year. For starters. With a capacity like that on the near horizon, it didn't seem too much of a stretch of the imagination to make O'Neill's vision possible.

Of course, the reality of the Shuttle program would be that it had much less ability to get materials into space than was promised, but O'Neill was not to know this. He proposed two main teams. The first was a group sent to the Moon to mine raw materials for constructing the habitat, making use of the Moon's much shallower gravity well to reduce the cost and effort of getting the materials into space (and at the same time holding back on frittering away Earth's overstretched resources). Once construction material was available in space, a second group of workers would begin assembling the first of the space habitats, envisaged as a pair of half-mile-long cylinders, tied together and set rotating around their mutual axis to generate artificial gravity.

SETTING UP THE HABITAT

The first space habitat, known as Model 1, was intended to house around ten thousand people, and O'Neill estimated that it would cost around $30 billion (this was based on the optimistically cheap costs of getting mass into space that NASA was predicting at that time for its as yet untried shuttle). When put alongside circa $20 billion spent on the Apollo program, this did not seem a financial impossibility. To O'Neill making it a reality was a no-brainer. Admittedly, the space habitat did not provide the same public spectacle as seeing the first person walk on the Moon, but it should have a much longer and more profound impact on the future of the planet and of human civilization. It wasn't a passing fancy, but a living future for thousands of people—a worthy enough goal and one that he felt was worth the investment.

O'Neill did not envisage his space colony orbiting the Earth

like a satellite. Not only would this give the habitat long-term stability problems, it would also have produced some thorny political
issues when governments were faced with the idea of a large, independent group of people and their technology (potentially including weapons) passing over their territory day after day. Instead, the
idea was to place the colony at one of the Lagrangian points, locations in space where the gravitational pull of the Earth, Sun, and
Moon largely balance out. Orbiting a Lagrangian is a much more
stable location than orbiting the Earth. Using the favored L5 point
would have put the space habitat in a rough equilateral triangular
position with respect to the Earth and the Moon, around the same
distance from each as we are from the Moon.

Most of the material—estimated at around 500,000 tons—would
have to come from the Moon, as it was simply too expensive to get it
off the Earth. Even from the Moon, rockets would not be financially
viable. But there was an alternative. O'Neill had seen articles about
the newly developed maglev trains that used electromagnets to accelerate a floating train, held aloft by a magnetic field. What's more,
he had himself worked on linear accelerators, devices that used
electromagnetism to accelerate charged particles up to high speed.
In theory a similar approach could be used to power building materials housed in metal canisters away from the surface of the Moon,
accelerating them to high speed before flinging them into the sky.

We have already encountered such electromagnetic catapults or
"mass drivers," the more romantic name O'Neill gave them. This
technology had featured in Arthur C. Clarke's science fiction, and
had even been the main mechanism for transporting materials
mined on the Moon to the Earth in another of O'Neill's favorites,
Robert Heinlein's masterpiece novel of a lunar colony's struggle for
independence, *The Moon Is a Harsh Mistress*. In O'Neill's version of
the technology a chain of metal buckets would be accelerated to the
Moon's escape velocity, flinging their contents into space before returning to the surface to be refilled.

O'Neill developed his concept, benefiting from input from veteran physicist Freeman Dyson who had continued to have an interest in spaceflight, and in 1974, O'Neill's ideas started to gain considerable public attention, making the front page of *The New York Times*, which with classic ability of the media to stretch scientific fact, took O'Neill's designs based on the very edge of what was possible with developments of the current technology and trumpeted that human space colonies were "Hailed by Scientists as Feasible Now."

O'Neill's ideas were a great match to many cultural themes of the day. Overlaying the traditional dream of expanding the frontier, was the opportunity to take these pioneering steps in a way that fitted with the new awareness of the fragile nature of the human environment on Earth. It is a common mistake to say that we need to save the planet. The Earth is going to be just fine, and there really is nothing we can do to it that it can't shrug off in a few million years. But keeping it as an environment where human beings can thrive is indeed a fragile thing, and the awareness of this was coming through in the new environmental movements of the time. Throw in a bit of science fiction razzmatazz, plus an appeal to the powerful desire to be freed of the constraints of big government—to get away from the ills of society as it had become and forge something new in a place unreachable by the old authorities—and it was clear that O'Neill's vision punched all the right buttons.

NASA TAKES THE PLUNGE

By 1975 even NASA was involved—or at least what Freeman Dyson referred to as the "paper NASA," the freewheeling ideas side of the administration that often seemed to have little connection to what actually got put into practice. A ten-week NASA summer study took O'Neill's plans and fleshed them out (at least as much as was possible), with engineers, physicists, and psychologists working together to produce far more detail on what could and should be done. One big change was to transform O'Neill's cylinders into something closer to a classic Clarke/von Braun space station concept—a huge, mile-wide wheel that would maximize the available living quarters for a given amount of materials.

The 1975 event was followed by two further NASA summer studies, building an impressive level of detail on what O'Neill had first proposed (even though he himself sometimes felt distressed that he was losing control of his baby). With such apparent enthusiasm making it seem likely that it would be put into practice, O'Neill gave some more thought to financing the venture. It was all very well to say it had a comparable cost to Apollo or the Space Shuttle program, but why would NASA and ultimately the U.S. government fund a venture for the benefits of those taking part, in what may well see itself as not so much a traditional colony as an independent state? Commercial funding has always played a major part of the extension of the frontier, while colonial states are rarely enthusiastic about their colonies becoming independent. You might as well ask the eighteenth-century British government how they felt about the Boston Tea Party.

PAYING FOR ISLANDS IN SPACE

While the basic idea of looking for commercial funding and reasons for the operation is sound—and we are seeing much more commercial-minded attitudes toward space in the twenty-first century—it is arguable that O'Neill got it wrong in settling on space-based power generation as the silver bullet that he felt would solve the financial problems of getting his space colony off the ground. The idea (see page 180 for more detail) is that it is easier to collect solar power in space than it is on the surface of the Earth, so you set up a solar collection station with vast banks of photo-electric collectors in space and beam the resultant energy down to the surface, using high power lasers, or their microwave equivalents, masers. However, to date the possibilities of making such a venture both financially viable and safe has beaten all comers.

If you had an optimistic frame of mind, it seemed entirely possible that these plans could start being put into action by the 1980s. Back in the mid-70s is was felt that by the time you are reading this book, what was sometimes referred to as Lagrangia would be long established and no doubt would be fostering several other new colonies of its own. Instead we haven't even seen a single person return to the Moon, let alone the launch of such a bold and impressive development. What went wrong? One early danger sign was a furious response to the plans from William Proxmire, U.S. senator for Wisconsin.

Proxmire lambasted NASA for wasting taxpayers' money on this fantasy. In reality, NASA had spent very little, but wary of negative publicity, and nervous as always of potential threats to the administration's budget, the NASA powers-that-be made sure that it rapidly backed away from any visible support for the concept. By the end of the 1970s there was also a fading of the counterculture surge that had made O'Neill's ideas seem so attractive to many. Rather than building toward a successful project, the seeds of decay had been sown and nothing was ever to come from all that careful planning and design work. It proved little more than a beautiful white elephant.

WALKING ON THE MOON

Even without the vision of mining materials to build O'Neill's space colonies, the first target to come into the sights of any colonizer is likely to be the Moon. It is the only celestial body that actually looks like landscape with the naked eye, and it has the surface area of a good-sized continent—enough to keep plenty of pioneers busy should they decide to settle there. It is also wonderfully near in interplanetary terms, a mere 380,000 kilometers (236,000 miles) away and taking around three days to reach. In terms of simple experience, the Moon is the only place apart from the Earth that we have ever visited—and we did it in the 1960s with technology that seems to have come out of the Stone Age when we look back on it today. There are washing machines now with more computing power than the Apollo capsule.

Some might see this previous experience as a negative. In 2010 President Barack Obama commented, "I just have to say it pretty bluntly here: we've been there before," as he made it clear that he had no intention of prioritizing a return to the Moon. President Obama was cancelling the Constellation Program, put together by George W. Bush to replace the shuttle with a more enterprising vehicle for reaching out into the solar system. Constellation combined the Ares launcher with its Orion ship (not to be confused with the 1960s nuclear-powered concept vessel Orion) that was envisaged as a vehicle for returning to the Moon. The president commented that the project cost too much, was behind schedule, and lacked innovation (familiar enough complaints, and all, arguably problems arising from the approach taken to fund NASA).

If we are to establish a sizable colony anywhere it is hugely beneficial if it has appropriate resources and raw materials on hand. The last thing we want to have to do is ship out more than the minimum from Earth, clawing our way expensively out of that deep gravitational well. There is good news and bad news as far as colonizing the Moon is concerned. First the good—there is plenty in the way of building materials like silicon dioxide and calcium oxide (the necessary components of cement and glass). There are metal ores to produce iron, titanium, and aluminum, and there's plenty of oxygen in those oxides too.

Just compare building structures on the Moon with the idea of constructing a colony in open space and you can see the huge advantages of having all these raw materials on hand. In space, everything you use (except sunlight) has to be shipped in from somewhere. On the Moon you are sitting on a vast sphere of available resources. In fact, as we have seen, even if you wanted to build a colony in space, you would probably get much of your material from the Moon rather than the Earth as it is so much easier to get mass off the surface.

HUNTING FOR SCARCE RESOURCES

On the other hand there are some distinct gaps in the riches that the Moon can provide. The most obvious limited essential is water. Water won't stay around as a liquid on the Moon's surface, frozen out of the Sun's reach and evaporating far too easily when hit by the light to remain accessible. It is possible that there is ice in some of the craters near the poles that are permanently shaded, but we don't know this for certain. In 1994 the probe *Clementine* (sent up as part of the Strategic Defense Initiative testing rather than by NASA) orbited the Moon searching for water using a wide range of sensors. There was no practical way to definitively spot water in these shaded locations, but using a form of radar it was reported that the results were "consistent with ice."

NASA followed up with *Lunar Prospector* in 1998. This had a neutron spectrometer on board, a piece of equipment that could detect hydrogen. When the element was spotted, it was assumed to be in the form of water at the Moon's poles at levels of around 0.5 percent. This is not exactly plentiful—it compares with some of the drier deserts on Earth—but it could be significant if that concentration represents infrequent large clumps of ice rather than tiny quantities distributed widely over a large area. Although NASA is reported to be developing a rover that can search for hydrogen, to combine with that plentiful oxygen in solid oxides to produce water, if the hydrogen is in any other form than water it would be necessary to work through a lot of surface material to get hold of a relatively small quantity.

While water consumption in a colony can be kept to a minimum by recycling at every opportunity, water is far too heavy to ship out from Earth and some of it will inevitably be lost. Also, for a real colony to be set up it must be possible to produce food, which requires both a steady supply of water and nutrients. The Moon has practically no nitrates or phosphates—essential to grow plants. What's more there is very little nitrogen, carbon, and (once again) hydrogen in any form, the building blocks that plants use to construct themselves. Again, a colony simply cannot afford to ship all these requirements up from Earth, and though there will be a certain amount available from waste products—space colonists could not afford to be squeamish about using their bodily waste as fertilizer—there would not be enough to use without supplementing it somehow. The same goes for the ubiquitous modern manufacturing materials for plastics, which require the same essential atoms as does life.

Then there's the matter of stuff to breathe. Not only does the Moon not have an atmosphere, it does not have enough gravity to hang onto one, even if we had some impressive way to pump the right kind of gases out. This has two significant impacts. One is that colonists would always have to produce their air—oxygen is easy enough, but nitrogen would be a significant problem. The other issue is the impact of the lack of atmosphere on solar radiation. The Earth's atmosphere is far more than something we breathe—it also acts a filter to block out a whole lot of the ionizing radiation from the Sun that would have a devastating effect in terms of generating cancers and making it impossible for plants to grow healthily. This means that any lunar colonies would have to use shielding domes that filter out the dangerous radiation, or would have to be built underground where rock can provide that shelter.

When it comes to growing plants—an essential for a permanent colony—there is likely to be a feast-or-famine situation with the incoming sunlight. We loosely refer to the "dark side of the Moon" as if one side is always in the light and the other in the dark. In fact both sides of the Moon get the same amount of sunlight, but we only ever see one side because the Moon's orbit is tidally locked into synchronicity with its orbit around the Earth. On the surface of the Moon, though, the experience would be a cycle of approximately two-week nights and two-week days. This is not going to make for easy agriculture and would probably limit growers to crops like algae, which would be more able to cope with these long periods of light and dark than conventional agricultural plants.

PRINTING A MOON BASE

If a permanent base were to be set up on the Moon it would clearly need the construction of appropriate buildings and this is one aspect of the venture where there is already an interesting proposal, though arguably in more of a paper NASA form than being real NASA yet. The architects Foster + Partners, responsible—among other things—for the new German Parliament building, Beijing Airport, the UK's Wembley Stadium, and the rebuilt World Trade Center in New York City, have produced designs for a Moon-based structure that could be constructed simply from onsite materials.

The ambitious idea is to transport an inflatable structure from Earth, which would then be covered with a shell based on regolith, the lunar soil that is composed of a mix of silicon, aluminum, calcium, iron, and magnesium oxides. This shell would be assembled using 3D printers, controlled by robots. The process has been tested in a vacuum chamber, using a huge "D-printer" capable of constructing house-sized objects. The initial design could house four individuals, though it could be expanded to become a lunar village over time. In principle the same approach could be used on Mars as well.

While we could envisage setting up an equivalent of the ISS on the Moon—a small-scale lunar base with regular visitors and supplies flown in—this is very different from a true colony in scale and reach. If a colony were to be established on the Moon, because it can never even come close to being self-sustaining on some life essentials, it would need in the long term to have an extremely valuable export that could counter the huge cost of shipping those necessities up from the Earth (or, for instance, intercepting comets to get hold of water).

A LUNAR ECONOMY

Just like building power stations in space (see page 180), there has been a suggestion that the Moon could export energy derived from sunlight. It is certainly true that a Moon base or colony could generate enough energy for its own needs relatively easily from the sun, though even this comes with a problem attached. Because of the two-week light, two-week dark cycle, it would be necessary to store enough energy to make it through the long lunar night. This couldn't be done practically with batteries, which are too heavy to transport and too inefficient to store enough capacity, and many of the ways used to temporarily store energy on the Earth, like producing hydrogen from water, or producing methanol from the air, are unavailable on the Moon. Probably the only option would be to have solar-collecting stations on both sides of the Moon—but that would mean bearing much larger infrastructure costs than a single site.

When it comes to generating energy to sell to the Earth, there certainly is plenty of lunar real estate on which to plant solar cells, and the solar energy they received would not be reduced by the atmosphere as happens on Earth, though the collectors would need some mechanism for tracking the Sun to maximize efficiency. However, the challenge of getting the energy back to the Earth becomes much greater than it would be from orbiting stations. And because the Moon is not in a geostationary orbit, it couldn't constantly be linked to a single site on the Earth.

Given that power station satellites in orbit around the Earth seem unlikely to be economic, the costs of generating power to send back home from the Moon (let alone of running the vastly more powerful microwave beams needed to cover the huge extra distance and dealing with the constantly moving target) would seem to rule this out as being totally infeasible.

It has been pointed out that the Moon has one scarce resource that will probably be worth the cost of extracting it and shipping it back to Earth in the future, although even this potential last resort comes with one proviso. The resource in question is the helium isotope, helium 3. Isotopes are variants of an element that have a different number of uncharged neutrons in the nucleus. This means that chemically they are identical to the most commonly found isotope, but their atoms are heavier or lighter, and can be more unstable, changing their effectiveness for purposes like nuclear power.

When the Manhattan Project was underway during the Second World War, one of the main problems was how to extract the much more unstable uranium 235 needed to produce a bomb from the dominant uranium 238. Similarly, helium 3 has the potential to be much more useful than the more common helium 4 in fusion reactors to produce electricity. Helium usually has two protons and two neutrons in its nucleus (the number 4 refers to the total number of protons and neutrons), but in the rare form helium 3, it has just a single neutron. This is potentially valuable for nuclear fusion reactors, which are likely to be among our main energy sources at some point in the future. Fusion reactors use the same types of mechanism as the Sun (though on a much smaller scale) and do not produce the same level of dangerous waste as our current fission reactors.

Helium 3 would be a great fuel for fusion reactors, but it is practically nonexistent on the Earth. All the evidence is that it is more readily available on the Moon (though even here it is only four parts per billion in the surface material). Helium 3 is such an effective source of energy that it may well be worth extracting it and shipping it back to the Earth, but the sheer bulk of surface material that would have to be processed and the cost of getting all the equipment in place to perform this processing are likely to mean the economics of producing it would only ever be marginal. Helium 3 is not the equivalent of "there being gold in them thar" lunar hills.

THE LUNAR SCIENCE STRATEGY

Taking together the requirements for energy and the essentials of life, the Moon does seem to have severe limitations when it comes to establishing a long-lasting and mostly self-sustaining colony that can keep itself viable by covering the cost of essential imports with high-value exports. It remains a good possibility for a scientific base, which doesn't have to be self-sustaining or financially viable—arguably if we can't put a base on the Moon we have no place attempting to head out any further. It would then be more like the Amundsen-Scott station at the South Pole on the Earth. No one suggests we colonize the Antarctic, or that we make it financially self-supporting. But the polar station has proved to be a valuable site to locate a base for undertaking scientific experiments. The same could go for the Moon, which, for instance, would be better than space in many respects as a location for telescopes.

The Moon enjoys the same benefits of lack of stray light and atmosphere as do satellite telescopes like the Hubble Space Telescope, but the lunar landscape could support a much larger structure. The Hubble has a 2.7-meter (106-inch) mirror—but the biggest telescope planned on the Earth at the time of writing, due to be built on Mauna Kea in Hawaii, has a 30-meter (1,181-inch) mirror. What's more, on the Moon it is possible to build arrays of telescopes that work together in a way that isn't practical with space telescopes, and on the far side of the Moon, a radio telescope would be shielded from Earth's messy emissions, making it easier to concentrate on the output of the stars. But apart from establishing scientific bases we need to look beyond our friendly neighborhood satellite, and, in something on the scale of the solar system, that means looking much further away.

THE PLANET FROM HELL

As we have seen, until the 1960s the obvious destination other than the Moon for a manned mission to set up a base or colony was thought to be Venus. This intriguing planet is the nearest thing the Earth has to a twin in the solar system, and apart from the Moon, it's our nearest neighbor, coming as close as 41 million kilometers (25.5 million miles), around 20 percent nearer than Mars at its closest. Venus is 95 percent the size of Earth and has a very comfortable 90 percent Earth gravity. Like Earth it is well equipped with atmosphere—in fact the thing that makes Venus so dramatic in the night sky, the next brightest object after the Moon, is its constant, thick reflective cloud cover. This made it impossible for telescopes to get any view of the planet's surface, but it seemed reasonable in the first half of the twentieth century that Venus would be like Earth but with a tropical twist.

Because Venus is closer to the Sun it was obviously going to be significantly hotter than we are used to—science fiction writers imagined a Venusian surface that was something like the Earth in an earlier, torrid, more humid period, perhaps covered with giant green plants and perhaps even occupied by higher life-forms. Venus, they thought, would be lush, tropical, and humid. A clear best bet for a colony. The reality came as something of a shock.

As we have seen, the first successful probe to approach Venus, *Mariner* 2 made a high pass at around 35,000 kilometers (22,000 miles) above the surface on December 14, 1962, and confirmed suspicions that had been aroused by a 1958 experiment studying emissions in the microwave region from Venus. The surface of the planet was hot. Not just tropically hot but at a temperature that would be disastrous for life—certainly making human exploration impossible. The average temperature has proved to be over 450 degrees Celsius (840 degrees Fahrenheit), hot enough for lead to be a liquid, peaking at around 600 degrees Celsius (1,100 degrees Fahrenheit).

This is hotter than the temperature on Mercury, which is far closer to the Sun. The contrast seems bizarre, but is the result of Venus having a very different atmosphere to our own. We are used to hearing concerns about the greenhouse effect that is caused by the Earth's 0.04 percent concentration of carbon dioxide. The atmosphere on Venus is over 96 percent carbon dioxide, providing a runaway greenhouse effect. To make the place even less hospitable, the atmosphere is so thick that the pressure on the surface is ninety-two times that on the Earth, and worse still, those clouds that give Venus its bright, white appearance are not water vapor, but a mix of choking sulfur dioxide and sulfuric acid, which descends on the surface, giving a whole new meaning to "acid rain." Venus is little short of a real-life hell and its environment makes the possibility of a colony there not worth contemplating.

Inevitably with Venus ruled out, thoughts turned to Mars—and it remains our best possibility for a long-term, large-scale colony in the solar system. Could the Red Planet, then, be our solar system's home away from home?

7.

THE RED PLANET

||||||||||||||||||||||||||||||||||||||||||||||||||

At most, terrestrial men fancied there might be other men upon Mars, perhaps inferior to themselves and ready to welcome a missionary enterprise. Yet, across the gulf of space, minds that are to our minds as ours are to those of the beasts that perish, intellects vast and cool and unsympathetic, regarded this earth with envious eyes, and slowly and surely drew their plans against us.

—*The War of the Worlds* (1898)

H. G. Wells

At the start of the 1964 movie *Robinson Crusoe on Mars,* an astronaut ejects over the planet as his ship runs out of fuel. His crewmate is lost, leaving him with only a space-suited monkey for company. The science in the film is terrible. The Martian air is breathable with a little oxygen supplementation, and radio messages from an Earth satellite arrive instantly. On the surface, fireballs present a constant danger. However, I'd still suggest this movie would be recommended watching for anyone volunteering for a trip to Mars, because despite being made in the early days of the real space program, it provides a gritty representation of the technical challenges of surviving in a remote, desolate location like Mars. As we will see there are real possibilities of attempts being made for a manned mission to Mars in the near future, but we should not underestimate just how difficult surviving there will be.

The Apollo missions took a little over three days to reach the Moon. At best, a mission to Mars is likely to take around six months—and because Mars and the Earth are in separate orbits, this distance varies wildly over a period of just over two years. At their closest, the planets can be as little as 50 million kilometers (31 million miles) apart, but more typically around 100 million kilometers (compared with the trivial 380,000 kilometers to the Moon), while when furthest apart they can be around 400 million kilometers away from each other—making careful timing of missions to Mars essential.

It has frequently been suggested in science fiction that the Moon could become a way station for constructing ships and putting together deep-space expeditions to Mars and beyond. As we have seen, this is possible in principle—the make or break seems to be just how much accessible water there is on the Moon. If there is plenty, this could provide air and fuel for the deep-space vessels. If the ships could primarily be constructed from lunar ore as well, it may be cost effective to stop off at the Moon on the way. Without these resources, all the materials would have to be hauled from the Earth, in which case a direct flight or a space-based platform would seem a more sensible starting point than the Moon.

THE RADIATION TRAP

As soon as we attempt to travel further than the Moon or to stay in space for a long time, a new hazard arises—radiation. This is even an issue for the occupants of the ISS, but it becomes significantly more dangerous on long-duration missions outside the atmosphere. The radiation in question is ionizing radiation, which is nothing more than high-energy light—X-rays and gamma rays—which cause damage to human cells. Medium doses significantly increase the risk of cancer, while high doses produce radiation sickness and death. The radiation can come directly from the Sun, but will also be produced by the impact of particles like cosmic rays (high-energy particles from the depths of space) and the solar wind. The effect is particularly strong during a solar storm, primarily from X-rays and gamma rays produced by the collision of these high-energy particles with the ship, collisions that give off energy in the form of ionizing radiation.

We tend to mentally link radiation to the artificial hazards of nuclear weapons and failing nuclear power stations, but it is a perfectly natural phenomenon. Even if our astronauts spent the entire journey to their destination in a thick, lead-lined box to protect them from the Sun and cosmic rays, they would receive some radiation, because human beings are themselves slightly radioactive, so emissions from one part of the body impact the rest. And when we are on Earth, there is always background radiation from cosmic rays and natural radioactive material on the planet's surface. Going out into space is not a matter of moving up from zero radiation exposure, but rather an increase in levels, with an accompanying increase in risk.

Again, even without venturing into space it is possible to change your levels of exposure. Every time you get on a plane, because there is less atmosphere between you and cosmic rays you get a higher exposure, around a hundred times the typical background and the equivalent of receiving a chest X-ray by simply crossing the Atlantic. Similarly, should you move from an average location to one with high background levels caused by presence of rocks like granite—say to Denver, Colorado, or to Cornwall in the UK—you will see your daily dosage jump by a factor of three. This only causes a small increase in risk, but for space travelers the exposure is considerably higher.

In space, cosmic rays and radiation from the Sun can exert their full damage without any screening from the Earth's atmosphere and magnetic field. The background radiation is already at significant levels due to the months of exposure arising from a mission to a destination like Mars, but there is also hugely increased risk should there be a solar flare during the voyage. These eruptions from the Sun's surface kick out vast quantities of particles, which in their turn can generate ionizing radiation on collision with the ship.

It is entirely possible to shield our astronauts from any radiation hazard—but shielding, typically using a material like lead, is heavy. As long as spaceships are assembled on the Earth and need expensive rocket propulsion to get them into space, there will always be a trade-off between expense, practicality, and safety. If we could assemble our craft in space, and use something like a space elevator to get materials up there things would be very different— but for the moment we have to accept that conventional shielding will be limited and must look at alternatives to keep our space pioneers safe.

One surprising but very practical suggestion is to make use of human excrement as a shield to protect the astronauts. Inspiration Mars Foundation (see page 149) has announced that both solid and liquid waste will be put in bags to use for radiation shielding after dehydration so that the water can be recycled. "It's a little queasy sounding," said Taber MacCallum of the Inspiration Mars team, "but there's no place for that material to go and it makes great radiation shielding."

It sounds as if this would result in insufficient shielding in the early part of the journey before sufficient mass has built up, but the expedition's food supply is also to be used for shielding, with the feces gradually replacing the food over time. The same thing would happen with the water supply, which would gradually be replaced by liquid waste. According to Marco Durante of the Technical University of Darmstadt, Germany, water makes a better radiation shield than metal, because it has a denser concentration of atomic nuclei, which block incoming particles.

One reason that the proposed Inspiration Mars departure (see page 149) is timed for 2018 is that not only is Mars closer to Earth than usual, it is also a low period in the Sun's eleven-year cycle of activity. In an intense solar flare even 3 meters (10 feet) of concrete would not necessarily be enough to stop the incoming radiation from reaching dangerous levels. The Inspiration Mars team has said that the astronauts would be traveling in the upper rocket stage of their vehicle and could use the length of the spacecraft as a shield by positioning it between them and the Sun, though this seems a worryingly ad-hoc solution.

THE BLACK FUNGUS OF LIFE

There is one unexpected plus side to the extra radiation in space, though—it might be used to grow a form of food. When a robot vehicle was sent into the remains of the Chernobyl nuclear reactor in Ukraine twenty years after the devastating explosion that wrecked the site, it came out with samples of a black fungus. The black coloration came from the pigment melanin, the same substance that we have in our skin to protect us from ultraviolet. Fungi have melanin for the same reason, but it seems that some species of fungus have developed a way to use the melanin rather as green plants make use of chlorophyll.

An ordinary green plant uses chlorophyll (and some back-end processing) to convert the electromagnetic energy of visible light into chemical energy, which then feeds the plant. But the black fungi from Chernobyl were feeding on the more powerful X-rays and gamma rays that make up ionizing radiation. This would make such fungi an ideal foodstuff that would both act as a barrier to absorb ionizing radiation, protecting the astronauts, and would grow extremely quickly in the high-radiation levels of space, providing a plentiful (if not very varied) food supply.

Although the dangers of radiation below a deadly level are always about statistical risk, the general feeling is that journeys of over 80 million kilometers (50 million miles) will be too dangerous without some kind of extra protection. At its closest, Mars is between 50 and 100 million kilometers (30 and 60 million miles), which means with a return trip (with the potential for extra time being exposed on the surface with no atmospheric protection) any Mars expedition is likely to require some form of protection for the astronauts.

An alternative to shielding is to reduce the astronauts' sensitivity to radiation. We have natural defenses in our skin, primarily there to protect us from the less powerful but still dangerous impact of ultraviolet light. The damaging effect of radiation is to break strands of DNA, potentially causing cell death or cancers. When radiation gets past the first defense of melanin, our bodies' ultraviolet protection mechanism detects breaks in the DNA and produces molecules called cytokines, which trigger the body's DNA repair mechanisms—but this process is localized and limited in scope, tending to be overwhelmed by the scale of damage that the exposure to ionizing radiation in space would cause.

NASA's Ames Research Center has a project underway to produce a mechanism that will artificially release cytokines when required. The easiest way to release a chemical internally this way is to use bacteria as carriers—but simply introducing these to the body is problematic, as the body's immune system will try to destroy the invading cells. Although the NASA team is yet to produce an artificially engineered bacterium that will provide the cytokines in response to ionizing radiation, it has found a way to get around the body's defenses.

The bacteria are enclosed in a tiny capsule, just 0.5 millimeter (0.02 inches) wide, made of carbon nanofibers. The shell prevents the body from detecting the presence of the bacteria, while its mesh-like structure keeps the bacteria in place but allows molecules the size of cytokines to pass through unhindered. If these capsules are placed under the skin, with appropriately programmed bacteria inside, they will release cytokines when exposed to radiation, an automatic defense that requires no monitoring or action from the outside.

A DIFFERENT WORLD

Even in simple semantic terms, when considering a space colony, Mars has a lot more going for it than the Moon. According to my dictionary, a "colony" is a body of people who settle in a new country or location unconnected to the state (or presumably in this case, the planet) they set off from. As we have seen, the Moon is only ever likely to provide the location for an outpost, a temporary site to visit and then come home. Outposts are certainly often manned by pioneers, but they don't really extend the frontier because of their impermanence. It is only with the founding of something at least semipermanent that we truly expand our horizon.

The *Apollo 11* veteran Buzz Aldrin, the second man to step on the Moon, is an outspoken enthusiast for founding a Mars colony. When asked why we should head out there he remarked, "Why did the pilgrims on the *Mayflower* set out to open up the New World? Because it's human nature to explore, to find a location to begin a settlement. And it is in reach." Not entirely surprisingly, the eighty-three-year-old former astronaut is more than a little impatient with our achievements to date. Considering the speed with which *Apollo 11* reached the Moon after Kennedy's initial declaration, he feels that our cautious, step-by-step approach to exploring of the surface of Mars with rovers has been far too slow, in part because of the lack of responsiveness due to the delays introduced by the forty-minute round-trip time for a signal to come from Mars to Earth and a response to be radioed back.

Aldrin quotes a program manager who had worked with Mars

rovers for five years, who believed that what had been achieved so far could have been done in a week if the human controllers were in orbit around Mars, rather than 50 to 100 million kilometers (31 to 62 million miles) distant. This might be an exaggeration, but there is no doubt that our remoteness does make each new step a slow struggle. Not surprisingly, Aldrin also promotes the pioneering spirit: "I think that the people who go [to Mars] will be remembered in history as pioneers, and the world leader who makes a commitment to establishing a permanent presence on another planet will also be remembered in history as a pioneer." A blatant attempt to encourage any listening world leaders to get started on a Mars program? Quite probably.

LIVING IN A BIOSPHERE

Much of the focus of planning a Mars expedition is inevitably on getting a crew there in one piece—an essential starting point, certainly. But if that expedition is to enable more than a flying visit, to build a true Martian colony, then the colonists have to be kept alive indefinitely in an enclosed environment, protected from the harsh conditions of the planet's surface. There is a lot that can be learned about how to do this (and perhaps even more to learn about how *not* to do it) from the experiences of the inhabitants of Biosphere 2, a sealed environment that was specifically intended to test the viability of an artificial enclosed ecosystem on another planet.

The name sounds as if it was the second in a series of experiments, but in fact it is a self-aggrandizing reference to itself as second, as the Earth is considered to be "Biosphere 1." Groups of eight and seven people were sealed up in a large, sprawling collection of sealed buildings that looked like a cross between a tropical hothouse and an architecturally challenging factory (much too large, at the size of two football fields, for any initial Mars colony). The first of the experiments ran from 1991 to 1993 and is often considered a failure. In the sealed environment, the participants could only live on the food they grew—something that proved to be an intensely difficult challenge—and had to use the water they recycled. One early lesson for the Mars venture was, "Be careful what you take with you." The experiment was plagued by swarms of cockroaches and ants, accidentally introduced into the biosphere with the soil and plant life.

The reason for the failure of the first "mission" was primarily a loss of oxygen. On Mars this could easily happen due to a leak, though on Earth any opening to the outside would have been an advantage, and certainly wasn't the problem for Biosphere 2. Nor was the loss directly due to living creatures consuming the oxygen, as that would have led to a corresponding increase in carbon dioxide levels, which did not occur. But the problem was the fault of living things. Bacteria in the soil were consuming the oxygen, but their carbon dioxide output was being absorbed by interaction with the concrete that was in close proximity to the soil in the relatively shallow growing beds. Without the carbon dioxide being released it could not be recycled to produce oxygen, making the air less breathable.

A key lesson for any real habitat for Mars was that Biosphere 2 made it clear that it wasn't sufficient to test individual components of the living environment in isolation. They had to be used together to see if any interactions would have unexpected results. This is an important consideration if Martian soil and rock is to be used in construction. Arguably the discoveries made in the experiment made it well worth undertaking. These are hardly lessons that anyone wants to learn when already on Mars, isolated from the Earth and unable to simply walk away safely like the Biosphere 2 inhabitants.

A second attempt was made to boot up Biosphere 2 in 1994, but this only lasted six months due to disputes over the management of the project. These became dramatic, including the removal of some management staff by federal marshals and sabotage of the environment by members of the first crew, which resulted in a breached air seal. Although the practical lessons from this run were less significant than the first, it still highlights another potential concern when such a mission is undertaken for real, particularly as space missions move to a more commercial basis. It reminds us that the controllers back on Earth can't order the pioneers to follow their instructions the way Mission Control might have done with NASA's astronauts. Part of the pioneering spirit is inevitably a desire to think and act independently. Any such operation needs a lot of thought put into the mechanism by which the volunteers make decisions and interact with their Earth support. They need to feel a degree of autonomy to be able to cope with a frontier environment.

Biosphere 2 was set up by a group known as the Synergists, led by John Allen. They were radical environmentalists, as much interested in gaining an understanding of the holistic nature of our environment as they were in experimenting with a suitable habitat for use on Mars. Based on this, some have criticized the Biosphere as a waste of time based on New Age drivel. But this attitude misses the real benefits of improved understanding. What was extremely valuable about the experiment, apart from the lessons from failure, is the way that the Biosphere 2 designers had to think through the implications of living in a sealed environment. For example, the structure had two "lungs"—huge circular sheets of rubber that could move outward as the air inside the structure expanded, to prevent the windows being blown out when the Sun was warming the environment. Similar consideration would have to be given in an equivalent Biosphere for Mars to the impact that the low pressure external atmosphere might have on the viability of traditional building design.

LIFE ON MARS

Where the Moon suffers from a water shortage, we know that there is plenty of water on Mars—and all the essentials needed to support life in the long term. Mars has frozen water at the poles, and there seems to be another source in the form of Epsom salt. This compound, originally made famous by the British spa town of Epsom and still popular as bath salt, is a hydrated form of magnesium sulfate, meaning that each molecule of magnesium sulfate has between one and eleven water molecules attached, which with an appropriate energy supply, can be liberated to produce fresh water. There is good evidence of deposits of Kieserite, the single hydrated form of Epsom salt, on Mars, plus the distinct possibility of a wider range of water-bearing crystals.

The Martian gravity is low at 0.38 times that of Earth, making a degree of bone density deterioration likely. However, what gravity there is, combined with appropriate exercise, should make Mars much better for long-term occupation than the microgravity of space stations. Even so, there may be need for supplements or drugs to keep bone mass at acceptable levels. The atmosphere is also very thin on Mars, currently only 0.6 percent of the Earth's surface pressure, and what little is present is not breathable as it is mostly carbon dioxide. However, it does reduce the impact of solar radiation to some degree, and the planet could in principle support a thicker atmosphere, as it once did in the past.

Lacking an atmosphere would mean that any agriculture would have to be under domes—and agriculture is essential as a colony needs to be able to feed itself. One of the key distinctions between the kind of outpost that could be established on the Moon and a Mars colony is that it is possible to imagine sending supplies to the Moon, but after an initial setup period, the distance to Mars is prohibitive when it comes to providing the essentials for life. If food is to be grown, it needs the water and shelter from solar flares already mentioned, but agriculture would also benefit from the fact that Mars has a rotation period only half an hour greater than that of the Earth, providing a steady, familiar progression of night and day, valuable for both plants to grow well and for human comfort.

THE NEW PILGRIM FATHERS

The distance involved in getting to Mars means that getting there is never going to be a particularly cheap process, though it is worth bearing in mind that this was also the case for the Pilgrim Fathers, whose journey across the Atlantic from Britain was neither trivial nor cheap. Though their journey took about half the time of a trip to Mars (and didn't have to contend with a lack of air or ionizing radiation, though in compensation the Pilgrim Fathers had to deal with a whole lot more weather), the cost has been estimated at $300,000 per person in modern terms. Given the current cost of getting mass into space and estimates for current technology to get someone to Mars you might be looking at around $30 million per person—but at the heart of this expenditure is the incredibly high cost of the particular form of energy being used to get into space.

Looking purely in terms of energy costs, we ought in principle to be able to bring down the per head cost of a trip to Mars to more like $10,000 if there was a different mechanism to escape the gravity well, but as yet that is an elusive possibility. We can see the costs of getting into space coming down, but not to everyday levels in the foreseeable future. And this means the costs of setting up a permanent base on the moon or a colony on Mars remain astronomical. Although, as we will see later in the chapter, commercial ventures are being proposed to visit Mars, it is unlikely that large-scale, permanent colonization will come from the private sector alone. The need for a financial return is natural and is too great.

We may feel that the frontier spirit is the absolute reverse of big government—that getting out to the frontier is all about taking personal responsibility and being a pioneer. And it is certainly true that this is a necessary trait in the individuals taking part in space exploration. But there is a big difference between someone crossing America and building a township from natural raw materials, which can be achieved without any significant financial backing, and the vast expense required to establish bases on the Moon and Mars. It is likely to remain beyond the capabilities of almost all businesses and will probably still require governmental input.

This may be the single biggest obstacle toward getting out there. At the time of writing we are in a worldwide recession and few governments are interested in multibillion-dollar projects, but even when money has been easy to come by, there has been a tendency to lump space exploration in with the science budget. This means there is a constant battle for funds, and where space gets money it is often at the expense of experiments that scientists would generally agree had much larger scientific benefit.

If we are to take on space properly without damaging very valuable scientific research, the two should be completely separated. Arguably, if space is to be bundled in with any budget it is defense, in part because this is one of the few budgetary lines of a government that are on the right scale, and in part because there is inevitably an overlap between the money put into development for space and for defense. It is much easier to envisage enabling real space exploration by taking a chunk from a worldwide defense budget, which was estimated to be around $1.75 trillion in 2012. Giving just 10 percent of this over to space projects would provide a healthy $175 billion, enough to enable dreams like proper bases on the Moon and Mars to be achieved. This would make much more sense than taking the cash from much smaller, hard-won science budgets.

THE MISSION TO NOWHERE

The immense distance also means a nontrivial journey time. As we have seen, the minimum crossing with the current chemical rockets is likely to be around 6 months, and a round trip passing around Mars without landing, even when our two planets are at their closest, is likely to take around 500 days. This was the reasoning behind the Russian Mars500 experiment, which began in June 2010 and finished in November 2011. Six international "astronauts" made the journey from nowhere to nowhere in a metal box in a Moscow car park. In effect this was "Biosphere 2, the sequel," not experimenting with the ways pioneers would survive on the planet but rather whether they would lose their grip on the journey through space.

The idea of the experiment was to simulate a return voyage to Mars and see how the six intrepid explorers could cope with living in a confined space together for so long. After a total of 520 days cooped up in a capsule, performing experiments and communicating with "the Earth" the astronauts emerged gaunt but apparently psychologically unimpaired by their experience. Everything was done to make things seem as realistic as possible. So, for, instance, when in contact with Mission Control and their loved ones on Earth, an increasing length of delay was put into the system, eventually at the midpoint a good twenty-minute gap was imposed to simulate the time taken for radio signals to travel between Mars and Earth.

Along the way, the Mars500 astronauts were put through stress tests simulating the kind of problems they could face on the voyage, including a total communications blackout. Each crew member was paid around $100,000 for their ordeal. Although the simulated astronauts were selected with as much care as the crew of a real mission, four of the individuals suffered significantly from a loss of productivity, insomnia, and a feeling of lethargy and lack of drive as they got into the tedious midsection of the long voyage.

While it is arguable that some of this lack of energy was because they knew they weren't going anywhere and that the whole "mission" was a sham, and so not worth fully engaging with their shipboard life; it is likely that real Martian pioneers, crammed together in a small vessel, will suffer similarly, in isolation with a very small group of humans housed all too close for comfort as their only direct contact. One crucial lesson from Mars500 seems to be that it is essential to give the Martian explorers or colonists enough space to be able to be alone if they wish to have time to themselves as this helps them avoid feeling too confined.

TAKING ON MARS DIRECT

There are already a number of plans in place that are intended to take a mission to Mars from being fiction to a viable project. One of the earliest to come within the realms of science, rather than science fiction, and still one of the widest in its scope is Mars Direct. This scenario, put forward by Robert Zubrin in his book *The Case for Mars* starts by sending an unmanned mission to the surface of Mars which carries a return vehicle, a nuclear reactor on a rover vehicle and 6 tons of liquid hydrogen. This hydrogen is reacted with atmospheric carbon dioxide to produce methane and water, which itself is split down to recycle the hydrogen and to store oxygen. The nuclear reactor provides the energy for this processing. The result would be to produce the fuel necessary to return to Earth, only needing to carry the hydrogen part of it out to Mars as a payload.

The second stage of the plan is to send out a four-man crew with a "habitation module" and three years' worth of supplies, plus a second, backup return vehicle, which, should all go well, would provide the return mechanism for the second manned launch. The plan is then to continue to send two vehicles every other year (fitting in with the dance of Earth and Mars around the Sun, so the distance flown is not excessive), one manned, one preparing for the next manned launch, to gradually build up a presence. Initially this would not strictly be a colony as the intention is to rotate the crews back on a regular basis—but over time the mission would be to build up a sufficiently large presence on Mars that the return capsules would become evacuation lifeboats in case of disaster rather than true return craft.

The most attractive aspect of the Mars Direct plan is that any particular mission would come in at a relatively low cost compared with a massive large-scale onslaught on Mars, yet the size of the base could build up to be significant over time. It is also a low-specification plan, not requiring much more advanced technology than was used in the Apollo program. The clever aspect of the "Direct" part with its homegrown fuel is that this approach does away with the old and complex model of using an orbiting return craft and a lander, as was required for the Apollo Moon landings. Having the two crafts would have been problematic for a Mars trip, as the crew left on the orbiter would be subject to high radiation levels and microgravity for an extensive period, meaning that the mission time on the surface would inevitably be limited. With the Mars Direct approach, missions can (in fact, would have to) last for 1.5 years, giving ample opportunity for exploration and scientific work.

BECOMING A VIABLE COLONY

Exploratory missions are one thing—full-scale colonization is another. In terms of getting enough people there to make a viable colony, the Mars Direct approach would, over time, be sufficient. If a long-term colony is intended, only a few of the launches would need to be emergency return vehicles—the rest could provide equipment and raw materials, but the fledgling colony would have to switch to using Martian resources as quickly as possible. The Mars Direct plan would provide oxygen and fuel, but there would be the need to be able to produce metals and plastics for construction and to assemble as large a living environment as is possible within the available time, apart from anything else to grow food.

It does not seem impossible to construct a colony on Mars that is self-sustaining in terms of food, raw materials, and basic manufactured goods, but there will be a need for more complex products to be shipped out to the colony for a considerable period of time after it has first been set up. It can be helpful to imagine that you were setting up a colony today on a newly discovered large island on Earth. It is easy to imagine that the settlers could soon be constructing for themselves the sort of technology that was available in the nineteenth century, or even producing basic metal or plastic objects with 3D printers. But it is hard to imagine them producing iPhones and hi-tech hospital equipment. They would need some items to be shipped in.

The Earth-based colony could get away to some degree with limited high-technology equipment, as they could always be evacuated, for instance in case of a medical emergency, but it is more likely to be essential to have the means for a safe modern lifestyle in the hostile environment of Mars. This high-technology cargo would need some kind of economic quid pro quo to make keeping the colony running worthwhile. In the short term—perhaps even for twenty or forty years—it is possible to imagine it being funded as a scientific project, or a government-backed venture for political one-upmanship, but to be a true colony there is a need to establish economic viability. It would be necessary to be able to financially justify the long-term existence and independence of the colony with exports that could pay for those high-technology imports until the colony was big enough to make its own.

EARNING A MARTIAN DOLLAR

Long-term Mars enthusiast Robert Zubrin has suggested that one of the exports that could help keep a Martian colony viable is intellectual property—the sale of ideas and information. In part this is almost inevitable in terms of media coverage. Whether adopting a reality show approach (see page 155), or traditional sale of rights, there is a lot of money to be raised from the likely enthusiasm back on Earth to keep up with what is happening on Mars. But Zubrin is thinking rather of marketing Martian ideas and inventiveness. And here I am not quite so convinced of the argument.

Zubrin's thinking is that labor shortages on Mars will drive a huge wave of technological innovation that will reap huge rewards—that Mars would, in effect, be an intellectual sweatshop, driven by its isolation. A typical Martian colonist imagined by Zubrin would need to be like Tom Sawyer in the eponymous Mark Twain novel, using his brains rather than brawn to get a fence painted. The colonists would have to be ingenious to get around all the problems they face with very little manpower, Zubrin argues. He uses the parallel of colonial and nineteenth-century America, but I suspect this has limited value as a model, both because an awful lot of the innovation that was happening in the early years of the United States was coming from Europe, not America, and also because there are plenty of examples of colonies that have focused all their energy on survival and have not been great centers of innovation.

It certainly is likely that there will be some creativity on Mars, but a lot of the high-tech, high-return innovation of the present day requires an infrastructure that simply won't be available to the colonists. It seems more likely that apart from the media aspect of intellectual property, a Mars colony would have to make use of extremely high-worth, rare, raw materials that are easier to come by on Mars than on Earth. These aren't abundant enough for it to be the driving force behind the establishment of a colony—a kind of "Mars Mining Company" ethos—but once the colony is established it could chip away at its economic independence thanks to these raw materials.

Another problem with the Zubrin model of the Martin economy, which sees Mars becoming rich as a result of all the patents generated by the smart, innovative colonists, is that while volunteers for the colonies are likely to be risk takers, and hence natural entrepreneurs (I am assuming that colonists will be volunteers, as Mars seems an excessively expensive destination as a prison colony—we aren't likely to see a parallel in the early years of Australia), these pioneers are less likely to be the really inventive geeks necessary to generate the flow of patents Zubrin envisages. The people who have the crazy but wonderful money-making ideas tend to be the kind of people who have no interest in taking a step out of the safety of their garage or study. Certainly not in being frontiersmen, constantly putting their lives at risk. I think it is possible to imagine a Martian colony that is economically self-supporting—but not one based primarily on its output of ingenious patents.

Sending raw materials, and low-tech processed goods, off-planet from Mars is certainly more attractive than it would be from Earth. The low-surface gravity makes it comparatively cheap to lift objects into orbit, or to a Martian moon, where some form of mass driver (see page 59) could be used to blast the cargo in the direction of Earth at low cost. (It still, of course, has to be gotten down to the Earth's surface safely without damaging the cargo or the cities below, a nontrivial and potentially expensive task.) The only proviso with this scenario being politically acceptable is whether or not those setting up the colony had read Robert Heinlein's classic science fiction novel *The Moon Is a Harsh Mistress*, which helped inspire Gerard O'Neill.

In the book, Heinlein envisages a lunar colony on the verge of independence from Earth. When Earth attempts to quell the uprising, one threat the Moon dwellers hold over their Earth counterparts is a result of the Moon's relatively weak gravitational well. The electromagnetic catapult that has been used to send minerals to Earth could equally be used to drop containers of rocks on Earth cities. These are big enough not to burn up entirely in the atmosphere and would hit the surface with the same destructive force as a massive nuclear device. If a colony gains the ability to ship large quantities of raw materials back to Earth this way, it also gains an awesome weapon. If Earth wants to set up such a colony, it had better be sure that it doesn't treat the colonists badly.

PLENTY OF ROOM OUTSIDE

When we consider Mars as a potential new frontier, it has a lot going for it. In terms of sheer scale, it dwarfs any new region that we have been able to explore before. Although Mars is significantly smaller than the Earth, because of the lack of oceans its surface area is very similar to the entire area of land available on our planet. By taking on Mars, we would effectively be doubling the land available for human habitation overnight. Mars has rich mineral resources and, as we have seen, a near-ideal rotational period. Admittedly it's a bit chilly with an average surface temperature of −63 degrees Celsius (−81 degrees Fahrenheit), but the maximum temperature can reach a respectable 35 degrees Celsius (95 degrees Fahrenheit) with that cool average primarily reflecting the lack of a decent atmosphere to act as a greenhouse effect blanket.

Along with the temperature issues, the lack of a breathable atmosphere is the other biggest obstacle to putting Mars on a par with a traditional Earth-based frontier. It is difficult to get past the outpost stage if it isn't possible to survive without expensive and potentially risky life-support technology. To move from an outpost to a long-term colony, the pioneers need to be able to relax a little and enjoy living, not just struggle to exist.

In fact it is arguable that a locale doesn't move from being a place for pioneers to explore to a true opened frontier where people can make their homes and build their lives until it is accessible without specialist gear. This is why attempts to portray the seas as a new frontier on the Earth have never really caught on. Yes, the oceans offer vast, untouched vistas compared with the often-crowded land, and they remain largely unexplored, but they don't provide a meaningful frontier because you can't live there as true colonists. The same will be true on Mars unless something can be done about the atmosphere.

ENGINEERING A NEW EARTH

In principle, the process, known as terraforming, is possible, although it is a vast engineering challenge and to date remains more in the scope of science fiction than practical engineering. It has been suggested that the best way to achieve terraforming on Mars would be first to thicken the atmosphere and increase the temperature by producing large quantities of greenhouse gases. This would have a positive feedback effect. As the surface warmed, it would release more greenhouse gases from the Martian soil into the newly thickening atmosphere. This would not result in air that was breathable, but it would enable humans to get around outside using simple respirators and would make it possible for plants to flourish in the open air. Crucially it would also provide a growing defense against ionizing radiation.

Such a process would be a long-scale undertaking. It could take fifty to a hundred years to get to the first stage of having a reasonably thick greenhouse gas atmosphere, and another thousand years or two before plants could generate enough oxygen to make the atmosphere breathable. It's not just a matter of getting oxygen levels high enough: We can't breathe if the carbon dioxide levels are too high, and initially these would be extremely elevated. It's a big task, but certainly not an impossible one. It is, after all, what happened on Earth billions of years ago. Our planet was initially oxygen-free. But the process here took place over millions of years, where on Mars it would have to be vastly accelerated by human intervention to work in a useful timescale.

Even though it would take *just* one or two thousand years this would be a massive undertaking with no historical parallels. It is hard to envisage taking on a program that has a similar timescale to that separating modern-day civilization from ancient Rome—and yet the scale of the task is equaled by the scale of the prize. If Mars could be terraformed, it would double our available space, relieving the pressure we put on our overloaded planet, and it would provide enough new frontier space to satisfy a whole army of colonists for ages to come. Although it is indeed a massive undertaking, there is nothing impossible about it, and with the right start, a colony on Mars could largely be achieved without shipping out resources from Earth.

TAKING THE PRIVATE ROUTE

With the legacy of NASA's Apollo success still echoing in our memories, it was unthinkable for most of the late twentieth century that a mission to Mars would be a commercial venture, but by the second decade of the twenty-first century it was becoming increasingly likely that the first attempt to put a human being close to that most desirable piece of real estate in the solar system would be a private expedition. Not everyone sees it that way, Canadian ISS veteran, Commander Chris Hadfield, responding to criticisms that the ISS has delivered little of value commented: "We will go to the Moon and we will go to Mars; we will go and see what asteroids and comets are made of. But we're not going to do it tomorrow and we're not going to do it because it titillates the nerve endings. We're going to do it because it's a natural human progression."

Hadfield's words seemed to be putting a damper on those who want quick manned missions to the Moon and beyond and seemed to be attacking the commercialization of the process: "We're not trying to make a front page every day and we're not planning on planting a flag every time we launch. That's just a false expectation of low-attention-span consumerism . . . It's just an uninformed lack of patience and lack of understanding of complexity and a desire to be amused and entertained that builds a false set of expectations." Hadfield argues that the ISS is valuable in providing lessons in construction and living in space that will be invaluable for future deep-space missions.

It is certainly true, as he suggests, that the ISS has proved a good testing ground for technology, with the advantage that it is possible to get back to Earth at short notice, being only 400 kilometers (250 miles) from the surface. A mission to Mars would not have this luxury. "We can [on the space station]," Hadfield said, "at any moment, when we have made a stupid mistake with a design, or an emergency that we hadn't recognized or because of human health, get in our spaceship and come home."

Perhaps, though, the views of this particular astronaut are a little old-fashioned. He clearly doesn't appreciate that "low-attention-span consumerism" is likely to produce most of the drive for conquest of deep space just as it got NASA to the Moon. And he also seems to be looking back to the Soviet-style massive state master plans for the conquest of space that were envisaged back in the 1960s. It certainly seems he is a little behind the times when his thoughts are put alongside the view of UK Astronomer Royal, Martin Rees: "[The ISS's] main [purpose] was to keep the manned space program alive and to learn how humans can live and work in space. And here again the most positive development in this area has been the advent of private companies which can develop technology and rockets more cheaply than NASA and its traditional contractors have done."

Despite the fact that at the time of writing President Obama was still supporting NASA's intent of sending a manned mission to an asteroid by 2025 and to Mars in the 2030s, presidential promises on space have rarely been met. This asteroid mission is certainly an interesting development, and paper NASA is at least engaging with the private sector in a novel way for this exercise. They have put out a call for ideas on how to capture an asteroid and put it into a lunar orbit—and had more than four hundred proposals from a mix of private and nonprofit organizations.

THE FIRST ASTEROID IN CAPTIVITY

The plan, officially titled the "Asteroid Initiative," involves capturing a 7-meter (23-foot) asteroid using a robotic probe with a "capture bag"—a metal-reinforced flexible container that is designed to envelop the asteroid to be able to then tow it into a lunar orbit using ion thrusters. The aim would be to achieve this within the next few years, taking a little under two years to reach the asteroid and another two to six years to get it into orbit. By then (around 2021) it is hoped that the Orion replacement for the shuttle (see page 195) will be available, so astronauts can go out and examine the asteroid at their leisure. It is indeed a manned expedition to an asteroid, through the mechanism of bringing the asteroid to our relative neighborhood means the humans don't have to travel anywhere near as far.

In some ways this is commendable. If the approach works it could become a regular "shopping trip," picking up highly valuable asteroids and towing them to an easily accessible range where they could be mined, providing raw materials for constructing deep-space ships or space habitats, or used (if it's the right kind of asteroid) to provide a fuel dump for future long-range ships. But there is a downside to this plan. By limiting the human travel to a jaunt from the Earth to lunar orbit, it means that this venture doesn't contribute anything directly to our ability to open up the frontier. Humans would still not be leaving our backyard.

Even this plan, though, depends on a major funding initiative. If the Orion program were postponed or cancelled, however effective the asteroid capture, there would be no mission to visit the Moon's new satellite. And there is little doubt that the financial climate following the world banking crisis is not good for major government space projects. Specifically the U.S. budgetary difficulties that led to the automatic reductions in 2013 is likely to delay any government plans, imposing around 9 percent reductions in NASA's budget. Perhaps the best hope for progress is NASA's newfound enthusiasm for working with private ventures—it has also put out a call for proposals for ways it can work with commercial Moon missions, ostensibly to share expertise, but perhaps also recognizing that they may have a better chance of getting there than Orion has of moving from paper to real NASA.

TAKING SOME MARTIAN INSPIRATION

So Lord Rees may well be right that the advent of commercial space travel is our best chance to build on what we have learned from the ISS, and it is encouraging that 2013 also brought not one, but two possibilities for commercial Mars missions to the fore. It was the first space tourist, Dennis Tito (see page 90), who hit the news initially with a plan to fly around Mars without landing. Tito's Inspiration Mars Foundation plans to send a two-man craft, a modified version of the SpaceX Dragon vehicle around Mars without a landing, making a relatively close flyby of the planet at around 160 kilometers (100 miles) above the surface.

Compared to the ISS, which orbits around 400 kilometers (250 miles) up, a 160-kilometer flyby would be a close skim. The mission is intended to take off on one of SpaceX's Falcon heavy-lifter rockets. These are currently still under development, planned to be operational around 2018, but SpaceX is not a paper company: it already provides the two-stage Falcon 9 which has become NASA's commercial solution to resupplying the ISS since the withdrawal of the shuttle. Inspiration Mars is intended to be a similar mission to the early Apollo flights around the Moon that demonstrated the practicality of what was until then a great unknown: Could human beings make such an epic journey through space and was the Moon an achievable target? This time around, though, the scale is much grander and the prize altogether more exciting.

What is most dramatic about Tito's plans is the timescale for the venture. As we have seen, missions to Mars are easiest when the planet is relatively close to Earth, and Inspiration Mars is hoping to make use of the window of high proximity in 2018, just five years after the plans were first announced. With its separate, larger orbit, Mars goes through cycles where it is further from and closer to the Earth. At opposition, its closest point in each cycle, which comes at intervals of a little over two years, Mars seems to do a loop in the sky. From the Earth's viewpoint it doubles back on itself, coming closest to the Earth for a brief period. But that cycle is not uniform.

Some oppositions are much closer than others. Although Earth and Mars come relatively close to each other every couple of years, the next chance for a similar close encounter is not until 2035, hence the urgency. (It's a shame we missed 2003, when Mars was at its closest for six thousand years.) Making use of the 2018 positioning of the planets will allow for a round-trip journey time of around five hundred days. Should the mission go to plan, the two-person crew—one man and one woman (probably a couple)—will be responsible for many human "firsts." Perhaps most importantly, this would be arguably the first space mission since the Apollo program that would really capture the public's imagination.

CLOSE ENCOUNTERS IN SPACE

There do seem to be two issues that should be learned from the experience of the Mars500 experiment (see page 137). That experiment involved six people, a number that makes it much less likely that there will be interpersonal issues than with a two-person crew. The Inspiration Mars plan to make the crew a couple is supposed to overcome this problem—and certainly there would be less concern about discovering unexpected interpersonal clashes. But there are plenty of married couples who might say that their worse nightmare would be to spend five hundred days unable to spend a minute away from their partners. The crew size on the Inspiration Mars flight is being limited to reduce the mass that needs to be carried—not just the weight of the individuals themselves, but the food, water, and oxygen they will consume. But this would be a false economy if greater numbers produced better stability.

The second potential problem is the decision that the Inspiration Mars team would be one man and one woman. In some circumstances this clearly could lead to problems. As it happens, the Inspiration Mars plan of using existing partners probably overcomes most such problems, but there certainly could be issues. The Mars500 experiment was single sex—all the earthbound cosmonauts were men. This was an intentional decision because an attempt at a similar kind of experiment in 1999 had to be cancelled early because a Russian astronaut tried to force a kiss on Canadian astronaut Judith Lapierre.

This earlier experiment was undertaken by the Institute for Biomedical Problems in Moscow and involved mostly Russian male astronauts and Lapierre. During a New Year's celebration, when vodka had been drunk rather more than might be advisable, the Russian commander took hold of Lapierre, pulled her out of view of the closed-circuit TV, and kissed her twice. The position was probably not helped by the response of Russian scientists monitoring the experiment. The coordinator Valery Gushin commented that Lapierre had "ruined the mission, the atmosphere, by refusing to be kissed." Though a single-sex crew would not necessarily eliminate such issues, it may statistically have improved the chances of an incident-free voyage.

People problems are only the beginning of the difficulties faced by Inspiration Mars, which has some major technical developments to make very quickly. As around three years is required in the build-up to the January 2018 launch, these problems would have to be overcome in about three years from the original announcement of the project, the kind of timescale for major technological innovation in space travel that has not been considered since the impetus given by Kennedy's speech.

AWAITING A HEAVY LIFTER

The first, and probably most nail-biting, of the issues is getting the Inspiration Mars probe on its way away from Earth. The Falcon heavy lift is a design from SpaceX, the company formed by the original owner of PayPal, Elon Musk. As we have seen, SpaceX does have a solid track record—its Falcon 9 launcher has already made the first private space launch into orbit in 2010, and in 2012 carried the company's unmanned Dragon capsule up to the ISS on a supplies run, which was repeated in 2013. But the Falcon Heavy, the launcher that is essential for Inspiration Mars, has yet to fly at the time of writing, with its first test due later in 2014. This makes it tight but possible for the Falcon Heavy to be ready for 2018.

If the Falcon Heavy were delayed, there is no opportunity to let the takeoff slip, as the window for making a practical trip to Mars is not huge. The only alternative to get the circa 10-ton Inspiration Mars capsule launched, with enough momentum to make its slingshot journey around Mars, would be NASA's promised Space Launch System designed for the Orion program, which is due to first fly in 2017. A year's leeway is very tight indeed for a project on this scale, which can easily slip or, in the case of NASA, be cancelled if government budgets are diverted elsewhere.

Taking off may be the biggest problem in terms of technology development and testing within a tight timescale, but landing the capsule on the return to Earth is likely to be just as challenging in a different way. Although there is doubt over the exact timescale of Falcon Heavy and the Space Launch System being tested and constructed, there is nothing truly novel about the technology being used. It should be pretty much business as usual, a very natural development from earlier manned space systems. But getting the Inspiration Mars craft back to Earth in one piece is much closer to the leading edge than the propulsion technology. One side effect of the slingshot around Mars is that the Inspiration Mars capsule will be traveling at higher speeds than any previous manned vessel, by a significant margin.

When the ship gets back home it is expected to spend ten whole days traveling around the Earth, gradually slowing with each orbit—but even then, as it comes down into the atmosphere, it will be traveling faster than any previous craft that has attempted a landing. At its peak speed, the space shuttle re-entered the atmosphere at around 30,000 kilometers per hour (18,000 miles per hour). By comparison, the Inspiration Mars capsule will be traveling at 50,000 kilometers per hour (30,000 miles per hour). More than ever before, the heat shield on the capsule that absorbs the heat generated as the craft enters the atmosphere and collides with air molecules, and the accuracy of its injection into the atmosphere, will be crucial for the survival of the astronauts.

STAYING ALIVE

The final requirement is to keep the couple alive throughout their journey. Much of the life support will be based on experience with the ISS, recycling waste water, and converting carbon dioxide back to oxygen; though less of this process will be automated, in part to keep weight down, to reduce the opportunity for things to go wrong, . . . and to give the crew more to do on their long journey. As we have already seen (page 127), shielding from radiation will also be an issue, though plans are in place to use food, water, and waste as a shield. Even with this in place, it has been estimated that there will be an increase in the risk of getting cancer by around 3 percent, and up to a 10-percent chance of experiencing a fatal level of exposure to flares of solar radiation called solar energetic particles, but the crew will have to accept this risk if they take part.

In fact, the entire venture—as is the case with Mars One, discussed below—relies on the fact that such a mission can be planned with a significantly higher failure risk than has recently been contemplated in a mission by NASA or any state-funded organization. To an extent this is a matter of transparency. The investigation into the *Challenger* shuttle disaster, in which Richard Feynman so dramatically demonstrated on live TV how the O-rings could have failed, demonstrated that the risks involved in NASA flights were known to be much higher than those that were publicly admitted to at the time—but Inspiration Mars would require a higher level of risk still.

At the time of the shuttle's withdrawal, and with the honesty that the *Challenger* inquiry inspired (or rather was forced on the powers-that-be thanks to Feynman's intervention), it was admitted that the predicted failure rate of the shuttle would be around 1 in 50 flights. Inspiration Mars is inevitably riskier, given all the firsts involved and the lack of an opportunity to make safe trials without human involvement. Though figures have not been published it is hard to believe that there would not be at least a 1 in 10 chance of a fatal outcome, should the mission go ahead.

Dennis Tito sees Inspiration Mars as very explicitly bringing back the frontier spirit to America. The Inspiration Mars website has the subtitle "A Mission for America," and puts the political and spiritual outcomes firmly alongside the scientific, proclaiming: "It will generate knowledge, experience and momentum for the next great era of space exploration. It will encourage and embolden all Americans to believe again, in doing the hard things that make our nation great, while inspiring the next generation of explorers to pursue their destiny through STEM education and exploration. Now is the time!" Tito seems to have grasped what many current politicians (and Chris Hadfield) have failed to see. Space exploration is not so much about science but about the political and social benefits and the boost in popular interest in science, technology, and engineering. It makes the subjects cool.

THE REALITY OF A TRIP TO MARS

Inspiration Mars aims to take its middle-aged married couple (it's hard to imagine anything more different from the classic NASA test pilot astronaut) on a close encounter with Mars, but not to land. However, the second of the commercial ventures with Mars in its sights, Mars One, has a bolder (and some might say a nightmare) vision of getting people onto the surface of the planet. There are two truly original aspects to the plan. One is to fund the expedition by making the whole venture into a reality TV show, producing the ultimate interplanetary version of *Big Brother*. It is interesting that Bas Lansdorp, the wind-energy entrepreneur behind Mars One, is Dutch, as was the original *Big Brother* format. With a pleasing circularity, it is also said that the original *Big Brother* TV show was inspired by the Biosphere 2 project (see page 131), itself designed in part as a model for a Mars colony.

The second innovation is that to keep things simple, Mars One only promises a one-way trip. In principle the astronauts might have to live out the rest of their lives on Mars, though the plans don't rule out return voyages being available at some point. Lansdorp is quoted as saying: "The technology to get humans to Mars and keep them alive there exists. The technology to bring humans from Mars back to Earth simply does not exist yet." This is probably an exaggeration, though the challenge of getting off the surface of Mars comes between that of getting off Earth and that of returning from the Moon. But it is fair to say that to achieve the timescale that Lansdorp has in mind, there is little chance of getting the Martian explorers back alive in the foreseeable future and it is better to plan on a lifetime on the surface of our neighboring planet as a likely outcome.

Like Inspiration Mars, the Mars One project is hoping to use SpaceX's Falcon Heavy and a modified version of its Dragon capsules. Unlike the versions of Dragon currently carrying supplies to the ISS, the Dragon landers would be wider and designed to become part of a habitat where a string of the landers are linked once on the ground. Mars One has a little more time than Inspiration Mars before it sends people into space, but not much, considering how much more there is to be pieced together in the project.

The aim is to have a first unmanned capsule landed on the surface of Mars in 2016. Assuming all goes well with what will inevitably be something of an experimental flight (and bearing in mind how many previous Martian probes have not made it in one piece), other capsules will be launched between 2018 and 2021, along with rovers which are planned to bring the capsules together, as no one can seriously expect to land them with such precision that they are close enough to each other to form an immediate structure. Mars One then intends to take its first four astronauts out in 2023 and another four in 2025.

RAISING THE CASH FOR MARS

Lansdorp thinks it will take around $6 billion to get his initial crew of four and their habitat onto the Martian surface, which he believes will be largely funded by the TV rights and the broadcast of the astronauts' lives, all the way from initial training through to the first walk on the Martian surface and beyond, presented as a reality TV show. Some have questioned the viability of this approach. Lansdorp is basing his figures on the fact that the International Olympic Committee raised $4 billion over around four years for coverage of the last Olympics. Chris Welch of the International Space University in Strasbourg, France, has commented: "We know how quickly the public lost interest in Apollo." Apparently by two years after the Apollo landing, public opinion of Moon landings had dropped by one-third from its peak. But I think Welch is wide of the mark.

It is certainly true that public interest did fade after *Apollo 11*— and apart from the surge of excitement with the *Apollo 13* rescue, it has never recovered. As we saw (see page 78) with my straw poll asking how many people knew who Charles Conrad and Alan Bean were, the interest (and certainly the memory) was already dropping with the second Apollo mission. But Mars One would be the first in a much more dramatic act of pioneering than reaching the Moon—and it would be coupled with good quality TV images compared with the terrible quality pictures we first saw from the Moon. Also Welch seems to overlook the impact of the hugely engaging (if ethically worrying) reality TV format.

I see no reason why it wouldn't be possible to win lasting audiences of sufficient size to raise the kind of funding Mars One requires. Lansdorp emphasizes that 600 million viewers worldwide tuned in to watch the first steps on the Moon, and Mars One could expect many more. I can see his point. I was one of those viewers in 1969, fourteen years old and allowed to stay up all night on my own—my parents had gone to bed, and the first step took place in the night for the UK-based audience—it was without doubt one of the most memorable moments of my life. We also have to remember that the potential worldwide TV audience is much bigger now than it was in 1969. Global events are far more truly global now.

"How many people do you think would want to watch the first humans arrive on Mars?" Lansdorp commented in an interview. "This will be one of the biggest events in human history. We are talking about creating a major media spectacle, much bigger than the moon landings or the Olympics, and with huge potential for revenues coming from TV rights and sponsorships." The comparison with the Olympics is apt. As we have seen, the Olympics typically generates $4 billion plus, and though a fair amount of this income is onsite from the live audience, around 75 percent typically comes from broadcast rights and sponsorship.

The Olympics might be a big draw, but it couldn't match the Mars landing for a peak audience, and certainly can't compete financially with a reality show that could be bringing in the dollars for fifty years. The lesson from Apollo is not that the fickle audience loses interest quickly. TV audiences come back for the likes of *Big Brother* year after year. It is rather that to be truly appealing, space missions have to emphasize the human aspect and to continue to be pioneering. While practically speaking there may well be a need for Mars Two and so on to build a more viable long-term colony, interest is likely to drop off and there may be a need for new missions to keep the model rolling. Asteroid One, perhaps; or even Jupiter One.

CHOOSING THE HERO OF THE FUTURE

In the most unnerving echo of *Big Brother* and similar reality shows, the plan is to let the TV audience decide which of the astronauts is to make his or her way into the history books and become the first human being to set foot on the surface of another planet. (No doubt another opportunity to raise cash as they pay to vote.) Although it is understandable why the reality TV approach is being taken from a commercial viewpoint, this does rather undermine the idea that the pioneers of this ultimate frontier are to be noble heroes.

It might seem that allowing the TV audience to decide on who makes that momentous first step is a kind of democracy. After all, why should the decision be made by a governmental organization like NASA or at the whim of a large, faceless corporation? Why not let the people decide? The problem with this view is that there is no democracy without free information. Reality shows like *Big Brother* or *American Idol* relentlessly manipulate the audience with the selection and presentation of what is shown. The choice of our first representative on another planet would not truly be in the hands of the public, but rather of the producers of the show—and that surely is not desirable. It veers too much toward the dystopian view of the movie *The Truman Show*—though at least the Mars One participants would know that they were in a reality show.

The whole Mars One setup also inevitably invites the question of what kind of person would volunteer for such a voyage. It isn't just the reality TV aspect, but the fact that it is a commitment for the rest of their lives with no provision made to get them back to Earth. Bas Lansdorp will certainly not be among the volunteers.

"When I was 20 years old," he told *The New York Times*, "and I first started dreaming of this, I thought I would go myself. But in the 15 years that have passed since, I know a lot more about myself. I am an entrepreneur. I am really not the right kind of person. I am impatient, and impatience is one of the worst character flaws for someone in a small group. [Also] I have a really nice girlfriend and I know she would never come with me."

Of course there will be a major selection effort involved in looking for what Nobel prize–winning Dutch physicist Gerard 't Hooft— who is associated with the project—has called "a near impossible combination of character traits." As you might expect, participants will have to be physically fit and mentally stable, able to live with others in close quarters, creative, adaptive, and able to take whatever is thrown at them. As the medical officer of the project, Norbert Kraft, has said: "If you take things too personally, you aren't the right person to go. If someone says, 'I need to climb mountains and smell flowers,' they are not the person for this . . . You should be able to survive in a hostile environment, and not freak out in a tin can."

The tin can in question is the space capsule, but the participants can't expect to be opening any cans of corned beef—or any other meat—either. The only option will be for the mission to be vegetarian until some form of animal-breeding program can be set up on Mars. (When it does, it is more likely to be more portable forms like chicken, rabbits, and guinea pigs than cattle.) Meat becomes a luxury when the weight of what is carried and the ease with which it can be produced in the harsh Martian environment are taken into consideration.

PICKING THE BEST

At the time of writing, the Mars One team claims to have over eight thousand volunteers, though clearly there will be a major sifting required to discover whether anyone is suitable. It is intended that between fifty and one hundred candidates will be chosen as the initial applicants, before whittling them down to the final handful of astronauts. This is a process that will include voting by the TV audience, though the Mars One program seems to be ensuring that their experts have the final decision in what is likely to be a crucial part of the mission. The intent is to select five crews, each two men and two women, who will train alongside each other for seven years until the public decides which of the four teams becomes the actual Mars One crew.

To give a better idea of how they will respond to the confined environment, the candidates are expected to spend three months of each year leading up to the mission in a simulated Mars base. By the time this process is complete it is hoped that the obviously dangerous candidates and those without a hope of success will be weeded out. It may seem a big commitment for the participants to spend three months a year in confinement, especially when you consider that most of them will not be in the final crew. But bear in mind that the ultimate "prize" is to spend the rest of your life in these conditions, with no way of getting out, so anyone finding the process unbearable automatically self-selects as inappropriate.

That 1 in 1,000 success rate for applicants seems optimistic in that the characteristics required may be even more rare among the population. More worrying still is the concern that the kind of people who would apply for a venture like this would be even less likely than the general public to be suitable. The suspicion has to be that most of those volunteering for such a challenge will be "unusual" personalities at best, and if the *Big Brother* show is any guide, this is an environment where interpersonal relations will be crucial. At risk of overstressing the point, unlike *Big Brother*, if things get too bad, contestants have no way of saying, "I want out. Let me go home." Perhaps the nearest equivalent the Mars One psychologists have to study are prisoners on a whole-life sentence.

But what stable individuals would volunteer for such a sentence, purely to be the first on another planet?

As with Inspiration Mars, at best the timescales of Mars One are incredibly tight. One concern is that the 2016 date is being adopted to get in ahead of the Inspiration Mars launch in 2018—because 2018 is a much better date for a launch. The 2018 date has been selected for Inspiration Mars because that is the point when Mars gets the closest it will be until the 2030s, at around 57 million kilometers (36 million miles). By comparison, in 2016 the distance to be covered will be 75 million kilometers (46 million miles), increasing the journey time by 30 percent. The manned Mars One missions in 2023 and 2025 would have to cover 82 million and 96 million kilometers (51 and 60 million miles), respectively. This means the second group of Martian colonists would have a journey time that is 68 percent longer than the lucky Inspiration Mars travelers. The next near-perfect opposition after 2018 won't be until 2035—too long a timescale for even these kinds of projects to contemplate.

NO PLACE LIKE HOME

Something that hasn't been mentioned much in the context of Mars One is that familiar bugbear of radiation. On the journey to Mars it is expected that similar precautions to those suggested for Inspiration Mars (see page 127) would be used. But what about on the surface of Mars itself? After all, these people will be exposed to raised levels of radiation for the rest of their lives. Tests from the Mars Radiation Environment Experiment carried on the *Odyssey* spacecraft showed that radiation levels in orbit around Mars were 2.5 times higher than those on the ISS. NASA estimates a maximum recommended exposure to these levels would be about three years.

On the Martian surface the radiation levels would be lower, but they are still a lot higher than is considered acceptable for a lifetime exposure on Earth. There are some regions on the surface of Mars with sufficient magnetic field to get some protection, but the Mars One pioneers will still be exposed to significantly more ionizing radiation than Earth dwellers—without the opportunity for cancer treatment should the worst arise. Even if it was practically possible, sunbathing would certainly not be an option.

The obvious possibility for protection from radiation is to go underground, but at the moment there is little evidence that this is part of the Mars One plans, which are based on a network of modules on the surface. Assuming the radiation levels on the Martian surface are typically similar to those on the ISS (so around 2.5 times less than in orbit), these are around 5 times the levels experienced on a plane flight, and over 300 times the average background level on the Earth. This will certainly increase the risk of cancer for the participants. The risk isn't as massive as some suggest—with eight people, there is around a 1 in 100 extra chance of someone getting a cancer each year. But it is a possibility.

INVADING THE MARTIAN MOON

If the difficulties of getting to Mars in one go prove simply too big to be practical and Mars One doesn't make it off the ground (or, worse, takes off but never arrives), one possibility would be to follow a suggestion made by former lunar astronaut Buzz Aldrin at the May 2013 Humans 2 Mars Summit. He thinks that rather than have humans attempt to assemble a base on the surface of the planet it would be better to house a small crew—perhaps three people—on the Martian moon Phobos for eighteen months.

Phobos is nothing like our Moon—just 22 kilometers (14 miles across) and looking much more like an irregular lump of rock, very similar in composition to some of the asteroids of the outer belt. It orbits much closer than our Moon too, just 9,500 kilometers (5,800 miles) above the Martian surface. Some think it may be a captured asteroid, though the chances of a passing body being captured and achieving a stable orbit are extremely low. Others have suggested that Phobos ended up orbiting Mars during the period shortly after the planets had coalesced in the early days of the solar system, or that it was the remnant of a collision of a larger body with Mars. Whatever its origins, it has practically no gravity—less than 1/1000th of Earth's—so it presents no problem for a ship approaching or leaving it.

While there, the crew would have clear sight of the Martian surface and would be able to remotely control the construction of the site without the twenty-minute delay faced by anyone attempting to do this from Earth. What may not have been thought through in detail is the combination of eighteen months of extra radiation exposure on the atmosphere-free moon, the impact of another year and a half with practically no gravity on the bones of the crew, and the fact that this would involve putting those three people through exactly the same stresses as the Mars500 crew—though the intense focus on the activity of remotely building a base might be enough to overcome the psychological impact.

SPACE PLAGUE

Whether we opt for a colony or an Apollo-style short-term exploration mission to the surface, once we land on Mars there is a new risk to be assessed that humanity has never before faced. Could exposure to material from Mars result in illness or death? This was not an issue with the Moon's dead environment, but it is entirely possible that there has been life on Mars and that some form of bacteria, viruses, or other dangerous or parasitic organisms still exist there, waiting to attack human life.

This may seem very much the stuff of science fiction, with plenty of novel and movie examples like *The Thing* where an alien infection transforms humans into monsters, but it is not a totally groundless concern. Some feel that anything sent back from Mars should be carefully quarantined, perhaps forever, at the ISS. There is even a group, the International Committee Against Mars Sample Return dedicated to preventing any material reaching Earth. Their argument is based on the panspermia idea, popularized by physicist Fred Hoyle, which suggested that life originated off the Earth, and thus extraterrestrial diseases could affect humans because we are all from the same source (some even believe that Earthly pandemics originate from space).

The alternative view is that this fear is unnecessary. After all, there is remarkably little cross-contagion between different Earth species considering how closely related all life-forms on Earth are. It hardly seems likely that an ancient Martian organism could have a close enough biological makeup to be able to harm a human. It is entirely possible that we may even discover that there never was life on Mars—we are yet to find any direct evidence of this—in which case the concerns seem exaggerated, to say the least. Nonetheless, should the Mars One plans come to fruition, the colonists would be foolish indeed not to take some basic precautions until it has been clearly established whether there is any potential for infection.

ENTER THE MARTIAN ENVIRONMENTALISTS

There is one other issue that the enthusiasts for establishing a Mars colony rarely seem to consider, which is that in return for a planet-sized frontier we would be effectively destroying a planet-sized area of wilderness. Some will regard this as ecological vandalism on a scale that makes our incursion into the Amazonian rainforest look trivial. They will ask the perfectly reasonable question of whether we have the right to despoil another planet, to make use of its resources for our own selfish benefit. After all, we've made a fair mess of the Earth.

There is also the consideration that we could contaminate Mars with our presence, by carrying bacteria that could depose any natural remnants clinging onto life there, for instance, or at the very least our contamination could confuse us into thinking we have found life on Mars when we haven't. At the moment, any interaction NASA makes with another planet is positively clinical, with a near-obsessive attempt to ensure no Earth material is brought to the surface of Mars. It really isn't likely that Mars One, for instance, could manage to take a similar approach given the level of sterilization required of current probes.

However, there have been recent arguments that we are being overprotective and that we are concentrating on the precautions that are easy to undertake, rather than thinking through what action has any real benefit. The existing levels of sterilization and protection make missions to Mars more expensive than they otherwise would be, and even with sterilization, our probes are likely to have carried some bacteria there already. However, it is highly unlikely, given the current Martian environment that we are going to see an invasive Earth species take over and displace the natives, as has happened when animals and plants have been moved from country to country, because Mars is not a particularly attractive environment for Earth life (as yet).

Similarly, it has been suggested that we worry too much about contamination from Earth giving false detection of life that will be assumed to be Martian, because it should be easy enough to make tests at a genetic level that would easily distinguish between a Martian native and an imported bug. In the end, once humans arrive there will be little that can be done to keep the surface pristine and arguably this particular battle will be over.

As for the arguments that we shouldn't be allowed to develop Mars, I think this too is probably not as great an issue as some would suggest. We would be coming at Mars with a much more enlightened viewpoint than humans have ever had in history. For the first time in taking on a new frontier we would be constantly aware of our impact on the environment, and operating in the harsh glare of international publicity, not living on the traditional isolated frontier with no oversight.

For that matter, there is too much of a tendency for conservationists (the clue is in the name) to want to set an environment in aspic. In reality, planets never stay the same for long. Bear in mind how different the Earth was during the glacial phase of the current ice age, just a few thousand years ago. All the evidence is that Mars once had significantly more of an atmosphere than it has now, and that it had liquid water on it surface. In some ways, terraforming Mars would be restoring the planet to an earlier glory.

It is also fair to say that we need to make a decision about which is more important—the future of humanity or the dead Martian environment. One of the main arguments for conservation on the Earth is that without thinking about environmental issues we could be damaging the future of human life, because we are highly dependent on the current, delicate ecosystem. But there is no equivalent to be concerned about on Mars, and to conserve the planet as the dead husk it currently is would be to preclude the future of life on its surface. Terraforming would indeed radically transform the Martian environment, but it would open up Mars for life (both human and other forms). And that is not a bad thing. Taking this radical step would raise a lot of legitimate concerns, but I believe that the opportunities far outweigh the possible negatives.

Apart from the potential benefits of the frontier, you could argue that without a terraformed Mars, living on Earth is like going on a cruise liner without lifeboats. In case of true disaster, there is nowhere to provide a bolt-hole. A terraformed Mars would not only give us a frontier and vast amounts of new space, it gives us a potential location to escape if Earth becomes uninhabitable. And we also should not forget that Mars, like other parts of the solar system, is rich in materials that we may soon be running out of on Earth. If we ever colonize the planet it is one possible workplace for space miners, a breed that has been ever-popular in science fiction.

8.

THE NEW GOLD RUSH

||||||||||||||||||||||||||||||||||||||||||||||||||||||||||||

And wild as it may sound, asteroids in particular could be highly profitable. In 1997 scientists speculated that a relatively small metallic asteroid with a diameter of 0.99 miles contains more than $20 trillion worth of industrial and precious metals.
—"How Asteroid Mining Could Add Trillions to the World Economy," *The Week* (2013)
John Aziz

There have always been pioneers who wanted to explore new territory out of sheer curiosity. It is a very human trait to want to know what's out there and to be the first to experience new locations. But it is equally human to feel the urge to discover and profit from new resources—to prospect for the riches that a new frontier can offer. The solar system is vastly well endowed in practically any natural resource we might desire. And fiction has often portrayed asteroid miners and other prospectors searching the skies for a heavenly El Dorado. But in reality we need to bring the harsh light of economics to play. What would we need to do to make it economically viable to gather resources from space?

At the moment, the economics of mining in space has to take into account the risk the activity poses to human miners. In the past, human life was treated quite cheaply, and as some suggest (see page 177) it may again in the future. But at the moment, making a journey into space has to put a cost on the risk to humans. So an inevitable question is whether mining can be done profitably without human involvement. We have already seen (see page 147) NASA's plans to capture a 500-ton asteroid and bring it into the Moon's orbit, and other automated ventures are planned.

MINING ON THE MOON

The Moon is a second target for the miners of space, specifically the chosen location for the start-up company Moon Express, brainchild of former Microsoft billionaire Naveen Jain. Moon Express aims to have a first mission underway in 2015. The Moon has the disadvantage of having a much deeper gravity well for would-be prospectors to cope with compared with an asteroid, but Jain believes that the Moon counters this with the advantage of having collected valuable material on its surface from asteroid impact for many eons. The Moon Express team also believes that the Moon's surface gravity will make the process much easier, as refining equipment, designed for use on Earth, would not need major modification to enable it to be used on the Moon, while it may have serious problems operating in zero g.

Clearly mining is not going to start on the Moon in 2015, even if Jain's company keeps to its fiercely tight schedule. All that is planned for that first trip is to prove that their lander can make a safe, soft descent onto the lunar surface and to do some early prospecting. The company expects to spend at least a decade on data gathering and prospecting before any serious extraction missions are sent out to the lunar surface.

One significant driver for Moon Express's thinking is the Google Lunar XPRIZE. In the same spirit as the Ansari X Prize (see page 91) that encouraged the development of *SpaceShipOne*, this is a major financial prize to help spur on competition. Building on the earlier Northrop Grumman Lunar Lander X Challenge, the Google Lunar XPRIZE is for the first privately funded team to safely land a robot on the surface of the Moon, have that robot travel 500 meters (0.3 miles) over the surface, and send images and video back to the Earth. There is a record-breaking prize pot of $30 million, making the Lunar XPRIZE the largest such incentive-driven competition of all time and inspiring over thirty teams to take part (though inevitably many will fall by the wayside).

ASTEROID MINING FOR BEGINNERS

Despite this lunar activity, the asteroids remain for many the obvious location for space mining, and this is the plan of Planetary Resources, a Washington-based company. They believe that it is feasible to mine near-Earth asteroids in the not-too distant future using robotic vehicles for everything from prospecting to extraction. Their plan begins with sending out a series of "Leo" Arkyd Series 100 space telescopes, which they describe as "a commercial space telescope within the reach of the private citizen." Although these telescopes are available for any purpose, Planetary Resources' own goal is to use the telescopes to scan near-Earth asteroids to see which are in range and most attractive for more detailed surveying.

The company has already signed up Virgin Galactic to provide its "payload services"—to get the telescopes into orbit—using Virgin's proposed LauncherOne booster, which is currently hoped to begin activity in 2016 with the aim of being able to deliver a 100 kilogram (220 pound) payload into low-Earth orbit for less than $10 million. This two-stage liquid-fueled rocket makes use of the same Virgin *WhiteKnight2* carrier aircraft as the *SpaceShipTwo* suborbital trip platform (see page 91), reducing the fuel requirement by launching from altitude, rather than ground level.

Planetary Resources claims that there are more than 1,500 asteroids that are close enough to Earth to take little more effort to reach than a trip to the Moon, without the hassle of transporting an asteroid like the NASA mission. Once the initial scan has been performed, the intention is to use a modified Leo craft, which has propulsion and additional instruments added to the original satellite. These "Interceptor" Arkyd Series 200 craft (it is perhaps a little worrying that the class names are so like something out of a video game) would undertake flybys to identify and track the prime-target asteroids. An even further extended version of the ship with laser communication ("Rendezvous Prospector" Arkyd Series 300) could also be deployed to take in more distant asteroids if required.

The final, key phase seems rather less well thought out than the initial prospecting. This would be to deploy asteroid mining robots that would extract the desired minerals and water. Some of these—rare metals, for example—might be dispatched back to Earth, while others—water is an obvious example—could be used to set up resource dumps in space to provide water, air, and fuel to manned expeditions exploring the solar system, meaning that they would not have to haul these resources all the way from Earth, potentially providing a major cost saving, or making it possible to carry far more in the way of useful payload.

THE ECONOMICS OF THE ASTEROIDS

I suspect that the potential for mining water is the more significant of the options. While at first sight the idea of mining precious metals from an asteroid where they are abundant seems so obvious (and is so reminiscent of a classic gold rush frontier), the economic realities of retrieving metals from asteroids are quite delicately balanced. It's true that, for example, platinum is plentiful in some asteroids and can currently demand an eye-watering $60 million for a single ton—but that value is dependent on its scarcity.

A useful image to put vast metallic asteroids containing a

plethora of precious metals into perspective is that it has been estimated that all the gold ever mined on the Earth would only be the size of a small office block, a lump about 20 meters (65 feet on each side). There has been far less platinum mined. It would not take much to flood the market. Business and economic pundit Tim Worstall has pointed out that when 250,000 ounces (around 7 tons) of platinum was once released for sale it was enough to drop the price by 25 percent. The more of these rare metals you bring in, the lower the price you will get for them.

Worstall asked Planetary Resources about this and their cofounder Eric Anderson responded by saying it was entirely possible that by flooding the market they would drive prices down by a factor of 100—but they would hope that sales might go up by a factor of 1,000, as with such cheap prices platinum would be more widely used in electrical applications where gold is the current preferred metal. But experts in mining finance find this a highly doubtful prospect. Apart from anything else, the new, expanded market could take a lot longer to develop than the timescale in which the prices would collapse—perhaps even tens of years. Prices could be devastated almost immediately on receipt of a huge supply. Economically, the business model of asteroid mining based on platinum and other rare metals is difficult to justify.

Putting together a cache of those essentials required for manned expeditions, setting up interplanetary rest stops or providing water for a Moon base, seems a more practical possibility economically as there will be demand for these resources, there would be a requirement for a lot more weight than there is likely to be for rare metals, and because it is hugely expensive to get water and other consumables off the surface of the Earth. As long as there is a base on the Moon or manned space program heading for destinations other than Mars there could be good economic sense to such a program—though the timing would be difficult to manage, as the caches need to be set up before the expedition takes place, yet manned expeditions are notorious for being cancelled (remember, for instance, George W. Bush's administration was taking us back to the Moon by 2012 to be followed by Mars). Running an asteroid mining operation to provide fuel, air, and water dumps would be a high-risk investment until manned missions had become commonplace.

KEEPING OUT THE OPPOSITION

One other issue is over ownership of the asteroids. As we have seen (see page 106) the Outer Space Treaty of 1967 probably doesn't prevent asteroid mining, but it does make such an operation a free-for-all. The wording is not entirely clear, but given there has not been any objection to selling space material that lands on the Earth in the form of meteorites it would probably be hard to argue that it was not possible to sell material from asteroids. And, as Eric Anderson of Planetary Resources has commented, it is unlikely that small asteroids under 500 meters (0.3 miles) across can be classified as "celestial bodies," the term used in the treaty for the locations that have any form of protection.

However, the treaty also makes it clear that you can't claim an asteroid for yourself, your country, or your company; it is the common property of humanity and everyone can come and get a piece of the action if they so wish. It has been likened to having access to the fish in the sea—they are not owned by anyone and up for grabs. But unlike asteroids, it isn't likely that two trawlers would argue over the same fish. Unless the treaty is renegotiated—and don't hold your breath for any such international treaty with less than decades to get it sorted out—it would be difficult to stop a rival from keeping an eye on your missions, waiting until you send in the miners, and then hurrying in to mop up some of the bounty without the initial expenditure on prospecting. At best we can expect multiple crews (human or robotic) on any asteroid, at worst we could see secretive sabotage and defensive maneuvers to keep out the opposition. This is not an ideal business environment, and certainly not for the fainthearted.

It seems the Outer Space Treaty would be useless to protect your operations commercially (some might envisage the kind of quasi-military space operations beloved of space opera, with different mining factions using weapons to defend their own miners against incursions from the opposition). However the treaty is also unlikely to be used in practice to prevent any activity. As commercial spaceflight lawyer Michael Gold points out, "The Outer Space Treaty is a tiger with no teeth. It's unenforceable and any state can pull out of it with a year's notice. I expect mining capability will trump the law in any situation." It seems unlikely this will be a major barrier to space mining.

A HOST OF FIREFLIES

Despite any concerns about practicality and legality, Planetary Resources is not the only company who feels that with the right vision and technology, automated asteroid mining could be made financially viable. Deep Space Industries is another company with ambitious plans to use a fleet of automated craft they have called *FireFly* (it isn't entirely clear if the name is based on the spacecraft in the Joss Whedon TV show of that name). Deep Space Industries (DSI) is headed up by Rick Tumlinson, who once worked for Gerard O'Neill and was a founding trustee of the XPRIZE organization. Although announced later than Planetary Resources, DSI has an ambitious timescale in mind, envisaging the first of the low mass (25 kilograms, or around 55 pounds) *FireFly* craft to be dispatched into space in 2015. (If even half of the predicted ventures take place, the years 2015 to 2018 are going to be very exciting for space missions.)

The commercial space business is littered with overambitious deadlines, and it could take a few years more, but there is certainly a significant resource backing DSI. Their intention is to move onto larger 32-kilogram (70-pound) DragonFly probes in 2016. Where FireFlies are intended purely as prospectors, the DragonFlies are expected to go on missions lasting up to four years, returning with at least their own weight in mined matter. Last in the current planned line is the Harvestor [sic] class asteroid collection devices, which are expected to be able to capture hundreds of tons of material a year.

Not all the material mined would be returned to Earth. DSI envisages much of it being used to provide fuel for space vehicles, or being used to construct devices in space. They hope to use 3D printers to produce products in situ, with a specific metal-wielding space construction device the MicroGravity Foundry planned to be in action by 2020. The MicroGravity Foundry would use lasers to draw patterns in a nickel-charged gas medium, depositing the nickel in precise patterns to produce intricate equipment, ideal for making parts of satellites, or, it has been suggested, as a repair kit for deep-space missions to replace failing parts when the astronauts are isolated from Earth.

As always, the economics of an operation like Deep Space Industries is open to question. When I queried the financial viability of their model, David Gump, CEO of DSI commented:

> We have two phases. In the prospecting phase, we sell data about asteroids to government space agencies at *far* lower cost than they are accustomed to paying (OSIRIS-REx is costing $1 billion), we sell payload space primarily to test out new components and systems that space agencies want to fly on a low-cost mission before trusting it on a high-cost mission, and we support corporate sponsorships and media events. In the retrieval and processing phase, we extract volatiles to extend the life of commsats (initially)

> with supplementary propulsion, and expand later to
> fueling in-space tugs for the LEO [Low Earth Orbit] to
> GEO [Geostationary Earth Orbit] orbit-raising service.
> We also extract metals for building habitats, tanks, solar
> arrays, spot beam antennas, etc., again for in-space
> markets where any mass (in GEO) is worth $17 million/ton
> due to launch costs.

Decoding this, it is certainly true that these missions would be a lot cheaper than the NASA asteroid mission OSIRIS-REx that Gump mentioned. The clumsily named OSIRIS-REx spacecraft is intended to launch in 2016 and travel to a near-Earth carbonaceous asteroid (101955) Bennu, study it in detail, and bring back a sample (at least 60 grams or 2.1 ounces) to Earth. DSI's plan is for DragonFlies to do this much more cheaply, though they still need customers with a budget for acquiring the data in what is always a volatile funding arena. The second phase envisaged by DSI clearly depends on a market that mostly doesn't exist yet. The difficulty here is what the timescale for that will be. There may not be much requirement for decades, which begs the question of whether now is the right time to develop such a business.

WHAT PRICE HUMAN LIFE?

The space economics that are essential to consider for space mining also has a dark side. It is one thing for individuals to take the risks that could enable them to strike it rich, but it is a very different matter when governments or large corporations decide they want to take over the action and put profits ahead of the well-being of the people on the front line. Literature professor Adam Roberts, makes a good point in his science fiction novel *Jack Glass*. The book is set in a future where there is a near-feudal hierarchy with a few controlling families at the top of society, a swath of corporations in the middle, and trillions of poor individuals scraping a living on subsistence at the bottom.

In such an environment, Roberts suggests, the balance between the three essentials for life of energy, resources, and human beings would be quite different to those that we think of today. As Roberts shows in his imaginary future, energy and resources are expensive, particularly to deploy in space—people are cheap and plentiful. So, for instance, in his imaginary world, robots exist, but are hardly ever used because people are a less costly and more easily replaced resource.

Our present economic approach and values suggest that it is better to send robotic probes into space rather than people, in part because the risk is high and we would rather lose a robot than a team of human beings. In the economics of Roberts's world this is inverted. You would always send people on a risky mission, because they are plentiful and easily replaced. It is possible that to be successful, a space-mining economy would require this kind of social inversion. While our initial inclination might be to produce hugely expensive automated mining equipment, this inevitably means that there is a high cost to be overcome before space mining becomes economically viable. In Roberts's world you would send out a ship full of human miners, because they didn't cost anything to speak of to manufacture, and were relatively disposable.

There are two issues with this view. The disposability of human beings is of course something that seems unpleasant. However, if we are honest, we don't actually put a high price on all human life, or there would not be the amount of suffering there currently is in the world. We are, of course, upset when human beings are killed in any disaster. But we inevitably tend to put a far higher weight on some people than others—particularly on the people who are close to us. The fact is that in the slums of a distant third world country, human life *is* cheap. I am not saying this is a good thing—it isn't.

But it is the reality of the economics of life on Earth and it is something that could, in principle be exploited in space by a ruthless corporation or government.

Even if we accept that some human life will be treated casually in space, though, there are some issues with Roberts's "people as a cheap and disposable resource" argument when it is examined closely. It makes great fiction, but isn't quite so good when looking at the practicalities of space travel, at least with existing technologies. While in terms of "manufacturing cost" people are cheap compared with sophisticated robotic equipment, and they are much lighter than robots that can carry out comparable tasks (hence requiring less energy to get them to the mining destination), people have different overheads from robots. A machine can get its energy from solar power or compact nuclear batteries, for instance, and can survive without air to breathe. People can't, needing large quantities of food, air, and water to be hauled along, which swings the economic balance back the other way.

In *Jack Glass*, the energy required to keep humans alive is kept cheap because the "sump," Roberts's slang term for the teeming masses (a contraction of "sub-polloi") exist on water and air mined from ice and a form of bioengineered food called "ghunk" that seems to be a variant of slime mold that can be grown pretty well anywhere as long as there is light, effectively making humans survive on energy derived from light. (In another Roberts novel, the poor have genetically modified hair that generates energy from photosynthesis to keep them alive, while the rich are conspicuously bald to demonstrate their lack of need for this.)

The easily deployed ghunk helps make Roberts's book a good story, but as yet such an approach isn't feasible in reality. It is entirely possible that we will develop such food technology, but without it, mining ships would have to carry a considerable weight of supplies, which would take us back to making humans more expensive than machines. Not because of the cost of risk, but rather the cost of providing them with energy and keeping them alive. Humans remain relatively fragile compared with machinery, so Roberts's vision is, perhaps thankfully, unlikely to come true anytime soon.

THE SPACE SCIENCE PARK

In the short term, setting up space laboratories or small-scale manufacture of very expensive goods are likely to provide an easier way to get initial financial return than getting miners working on natural resources from planets, moons, and asteroids. Although the ISS has not been overwhelming in the valuable research it has generated, there has been some scientific work undertaken, and it is arguable that once commercial research capacity is available in space it would be more attractive than the ISS, as any commercial ventures where large sums are to be spent on research are likely to be secretive, an impossibility in the shared environment of the ISS.

There is no doubt, in principle, that interesting work could be done making use of the low gravity, the easily maintained low temperatures and the high vacuum available in space, whether it is at the level of experimental electronics or in small-scale trials of specialist manufacturing. It is certainly a lot more practical to generate product ideas and trials with their associated patents in a space-based lab that it is to manufacture on a large scale, with all the difficulties of getting raw materials into place, shipping the finished goods to Earth, and keeping costs down in what usually has to be a relatively low-cost operation to make mass production viable.

Another alternative money-spinner might seem to be in building space hotels. As we have seen, a few very rich people have already been prepared to pay millions of dollars to get into space. There are plenty of larger-scale ventures planned, from the glossy Virgin Galactic to more speculative start-ups, but as we have seen, most of these future-gazing companies are only offering the opportunity to skim the edge of space for a few minutes or hours, rather than take a trip to some form of space station.

Although space hotels have been envisaged for decades, the cost of being a guest would have to be in the $1 million to $10 million range, which severely limits the number of potential visitors and makes the economics of running such a venture doubtful at best. Realistically to be viable such a space hotel would have to find a hundred or more guests a year—quite a challenge. And for such a venture the safety standards are likely to be set considerably higher than for a true pioneering exploration—causing the already high costs to inflate dramatically.

POWER FROM THE SKIES

Alongside tourism there is another commonly suggested way to make money out of space that seems on the surface to be attractive. This is to beam power from space; in effect to mine raw energy, which as we have seen, was suggested by Gerard O'Neill as the way to finance his space colony concept. If there is one commodity that will be in short supply in the next few decades it is energy, especially if we heed the calls to reduce our reliance on fossil fuels with their heavy burden of greenhouse gases. Solar power is attractive, but harvesting it is restricted on the Earth's surface by the space the collectors take up, by interference with incoming sunlight by the atmosphere and weather, and by the inevitability that in any location, sunlight collectors will spend a fair part of the day with the Sun out of line of sight.

Why not instead place vast arrays of solar panels in space, where there is no atmosphere to weaken the incoming solar energy, with the potential to track the Sun 24/7, then beam the energy produced down to Earth? It sounds like an environmentalist's dream, but there are some serious practical problems attached. Such a satellite would probably have to be in a geosynchronous orbit, so that it can keep in line of sight with the Earth station that receives its beamed energy—but that means placing it in a high orbit that is expensive to achieve. It is one thing to do this with a tiny TV or communication satellite—quite another with a vast solar farm that would take many rocket launches to get into space.

It has been estimated that launching a big enough solar power satellite station to viably produce energy would cost trillions of dollars—vastly more expensive than any land-based technology to produce a similar amount of energy—and though we can see launch costs coming down significantly in the next decades it is highly unlikely they will drop sufficiently to make this a cost-effective way of generating energy. These problems were recognized when Gerard O'Neill proposed the idea. O'Neill suggested that they could be overcome by constructing the solar power systems on the Moon and transporting them to the appropriate Earth orbit from there, vastly reducing the cost of getting the equipment into space as a benefit of the Moon's much shallower gravitational well.

If we had a large-scale Moon base already established—with the infrastructure required to mine raw materials, manufacture the solar panels, and send them into space—then it is likely to be true that this would be a cost-effective solution. But if, as seems more likely, the Moon base would have to be built specifically for the purposes of producing the solar power systems, the initial setup cost of that base with a suitable manufacturing capability would dwarf that of getting a system into place from the Earth. It would still be a hugely expensive way of producing energy, unless we had other good reasons to set up that base.

Finally there is the nontrivial issue of getting the energy back to Earth. Even if you had a space elevator (see page 61), the elevator's cable would be far too fragile to carry the kind of power that such a station would generate. Such a narrow ribbon may be able to carry enough energy to run the elevator itself—the carbon nano-tubes that are often spoken of as a potential material to make the ribbon because they are so strong are also excellent conductors—but the quantity of energy that would have to be sent down is immense. There is no point setting up a hugely expensive space power station to generate a few megawatts. The only potential mechanism would be to use a microwave beam.

Ever since Nikola Tesla's work at the end of the nineteenth century, engineers have dreamed of being able to beam or broadcast energy wirelessly. Tesla believed he had the mechanism to do this that was based on sending a wave through the Earth, but all the evidence is that he was deluded. (Tesla was a superb engineer, devising the AC electrical system, but his physics was never great; for example, he never got the hang of relativity, convinced that Albert Einstein had it wrong.) It is possible to send energy through the air relatively efficiently using microwaves, with as much as 50 percent of the energy being available at the far end of the process. But there are very good reasons why we don't have microwave beams acting as national energy grids, instead opting for low technology with messy looking, easily damaged cables and pylons. A high power beam of microwaves is not something anything living wants to encounter.

Microwave ovens work by producing vibrations in electrically polarized molecules like water and fats that are present in foods, causing them to heat up and cook. A vastly more powerful microwave energy transmission beam would have exactly the same potential for cooking anything living that strayed into its path. In Tesla's time they might not have worried too much about the safety issues involved, but today it is hard to envisage it ever being allowed. Given the fuss originally caused by birds killed by wind turbines (typically a couple of birds per turbine per year), a constant stream of dead wildlife dropping out of the sky would not be popular—let alone the risk of instantly fried passengers if any aviation came near the microwave beam. And at some point the beam from the satellite has to reach the ground—but a tiny shift in the satellite producing it could result in the beam drifting miles out on the surface, destroying anything living in its path.

MINERS ALWAYS TAKE RISKS

In the end, the space miner remains one of the few likely ways to undertake commercial activities in space. Unlike tourism, a lot of the income can come from automated devices, only needing relatively sparse expensive human input (assuming Roberts's inverted economy does not come into force) for prospecting and acting as supervisors of robotic mining devices. Like it or not, as a society we are happier to accept relatively high-risk environments for specialists like miners than for the population as a whole (particularly tourists). Compare the media storm that blew up over the problems at the Fukushima nuclear power plant in Japan in 2011 following the earthquake and tsunami, with the kind of coverage that is given when groups of miners are killed in accidents.

At least sixteen people were killed in a gold mine accident in Ghana in April 2013. In the same month, eighteen people were killed in a coal-mine explosion in China. These events hardly made the news. Yet the Fukushima plant disaster, headlining news around the world, killed no one. Part of this disproportionate response is a result of the extreme level of fear that nuclear radiation raises, stirred up by the way the media treats such accidents, but it is also a reflection of the way that we accept a significantly higher level of risk for workers undertaking a job like mining (especially if they are far away) than we do for risks to the general public. The chances are that space mining will be able to keep costs down by having safety levels significantly below anything that would be required for a space hotel.

The risks involved in reaching Mars were covered in the previous chapter, but all the same problems turn up in a more extreme form if the target is the asteroid belt. Situated between Mars and Jupiter, as we have seen, the belt was long considered to be the remains of a planet that exploded. This was supported by the unscientific but popular idea of "Bode's law," which posits a pattern for the location of planets around the Sun. The "law" predicts that there should be a planet where the asteroid belt now resides. However as we have gained a better picture of just how much material there is in the belt it has become obvious that there was never a planet there. There is far too little material—adding up to around 5 percent of the mass of the Moon.

A WHOLE LOT OF SPACE

The image of an asteroid belt that we get from the movies, or from computer games, is that these floating lumps of rock, metal, or ice are crammed together, making navigation through an asteroid belt a dangerous maneuver (but a fun gaming activity). So much so, in fact, that *Star Trek*'s Captain Kirk and others have been known to dip into an asteroid belt as a kind of interplanetary briar patch where superior piloting skills make it possible for our heroes to hide from their (much less skillful) aggressors. The real asteroid belt, at least in our solar system is very different. Yes, there are thousands and thousands of asteroids—but the space they inhabit is vast.

For a ship passing through the belt the experience would be very different from the frantic ducking and diving to avoid collision that we expect from the movies. The chances are high that such a ship would never have to change course on its way through the belt because it is unlikely to directly encounter a single asteroid without making the effort to do so. Should it, by chance, arrive at an asteroid, in most cases there won't be another one in sight anywhere around it—they are that far apart.

The modern conception of the asteroid belt is that it is not the result of an exploding planet, but the leftover debris from planet formation. When the solar system was young, the planets condensed from a spinning disk of dust and rocks—the asteroid belt is the remaining rubble that wasn't close enough to a planet to have become a part of it. However, from the viewpoint of a miner, the asteroid belt is literally and metaphorically a gold mine. It has a wide range of attractive and rare ores, plus a fair amount of frozen water.

It has been estimated that a typical, relatively small 1-kilometer (0.6-mile)-sized asteroid will contain around 200 million tons of iron, 30 million tons of nickel, 1.6 million tons of cobalt, and enough precious metals to have values in the hundreds of billions of dollars (subject to the dangers of flooding the market mentioned on page 171). And there are thousands of such asteroids. If you can get there, it is much easier to send your mined materials back to Earth as there is no gravity well to overcome. The only problem is that out in space there are none of the home comforts that even the barren surface of Mars can offer.

To reach the asteroid belt would not necessarily take much longer than getting to Mars—typically around nine months each way—but the entire journey, plus any time spent during the mining process would involve being exposed to solar and cosmic ray radiation without any atmospheric or magnetic shield, and would mean spending an impractically long time without gravity, which would result in bone and muscle deterioration.

Some of the shielding ideas suggested for a Martian trip (see page 127) would obviously help with the radiation problem, and in practice the risk of the kind of exposure expected on a two-year round trip (around 100 rem), assuming no major solar flares are experienced would probably be tolerable, bearing in mind the overall risks involved. This is around one thousand times the recommended annual limit for exposure to artificial sources, and would result in around a 5.5 percent increase in risk of cancer.

PUMPING UP THE GRAVITY

The more immediate problem would be from being in space for a sufficiently long time for the low gravity to have deleterious effects on the body. For a journey of this length, some kind of artificial gravity would be required to keep the astronauts healthy. We know from the equivalence principle, Einstein's observation that triggered his thinking on general relativity and the nature of gravitation, that acceleration and gravity are indistinguishable with certain provisos. One possibility would be to constantly accelerate the spacecraft under power.

The good side to a constant thrust is that it would give an apparent gravitational pull toward the rear of the craft, felt throughout the vessel. (The gravitational effect is felt in the opposite direction to the acceleration.) And it would result in the ship reaching its destination much faster than a typical "accelerate then drift" approach—but there is good reason that spacecraft don't use their engines for a long time. Keeping a ship under power takes an awful lot of energy. Unless it could be accelerated by a combination, perhaps of solar sails and a nuclear engine, there is no way that sufficient fuel could be carried. There is also the problem of how the miners would maintain gravity when they arrived at the asteroids.

A more practical approach to generate artificial gravity, much favored in science fiction, which we have already seen, is to spin the ship. It might seem at first sight this lacks the appropriate input of acceleration to be indistinguishable from gravity, but the acceleration is still there. Think of the kind of carnival ride that features a large, upright metal cylinder. The riders stand with their backs against the inside of the cylinder, which is then spun around. At this point the floor of the cylinder drops away, but the riders stay stuck to the cylinder—the more daring can even turn upside down in position as long as the cylinder keeps spinning.

What is happening here is often described as "centrifugal force," flinging the riders outward—but this is a misrepresentation of reality. Think what would happen if the cylinder suddenly disappeared. The riders would shoot off in a straight line that was at a tangent to the cylinder wall. But put the wall in place and the metal applies a force to the rider pushing them inward, accelerating them toward the center. This inward acceleration results in the equivalent of gravity heading outward on the cylinder, pinning the rider to the wall.

So in principle all you need to do to generate artificial gravity is to spin your capsule. It would be different from the constant acceleration ship, as the amount of gravity would vary depending on where you are in the vessel. Someone in the middle of the spaceship near the axis of rotation would not experience gravity—it would only be felt at maximum close to the outer wall. This isn't too much of an issue, and could even be an advantage for undertaking low-gravity work during the flight. But there is also a more significant problem.

We have a mechanism inside our head, based around the inner ear, used to detect acceleration. It's not unlike the accelerometer in a cell phone that can detect how the phone is twisted and turned. This system provides the brain with data to help keep the body in balance and make us aware of our orientation. But should you subject someone to constant rotational acceleration, the system gets confused by the forces that results from rotation—the person under acceleration becomes dizzy and disoriented and suffers from motion sickness.

Most of us have felt this on a rotating fairground ride—imagine a ride that you are forced to stay on for several years at a time. With a big enough rotating object this is no longer an issue, as the effect is less pronounced for the same level of artificial gravity. With a mile-wide "wheel"-shaped space station, for instance, the wheel would only have to spin once a minute to deliver Earth surface gravity, a rotational speed that would not induce any motion sickness. But this does mean that a conventional rocket-shaped cylinder of a ship design is not likely to be acceptable for the long-distance traveler.

Given that such a ship could spend its working life in space without landing or taking off under any significant gravitational field there are several designs that could provide the right effect. It would be possible, for instance, to have a large toroidal (donut shaped) section in the middle of the ship that rotates and provides the living quarters. Other designs have placed the living quarters in a tumbling cylinder, more like a traditional centrifuge. The amount of artificial gravity generated goes up with the distance from the center of rotation and with the square of the speed of rotation. It should be possible with speeds in the 4 to 6 revolutions per minute range or less, which do not cause excessive discomfort on the scale of a reasonable-sized ship, to generate sufficient gravity to hold off bone deterioration.

THE TRIANGULAR TRADE

Taken in isolation, while the asteroids offer rich pickings, getting to them from the Earth remains a major expedition—and they are unsuitable for a permanent colony. Not only is there the lack of gravity, they are too far from the Sun to get enough energy, while any particular asteroid will typically lack the right mix of material to form a base (usually an asteroid will either be metal rich or carbon/water rich, but not both). The most likely scenario for asteroid mining to become successful is if a Mars colony is already established, as reaching the asteroids from Mars is much more cost effective than getting there from Earth, both because of the much shorter journey time, and because of the shallower gravity well to get away from the planet. Robert Zubrin has calculated that it would take fifty times as much mass of fuel, storage tanks, and so forth to get a ship to a main belt asteroid from Earth than it would to get the same ship there from Mars—a huge saving factor.

The imbalance suggests that with a Martian colony established, as asteroid mining comes onstream we could see a classical triangular trade route building up—the process that was well established in British colonial days between the UK, North America, and the West Indies. In this parallel, the Earth is Britain, North America is Mars, and the West Indies is the asteroid belt. Manufactured goods requiring high technology would still flow from Earth to Mars, where the premium on specialist manufacture would cover the expense. Lower-technology manufacturing would take place on Mars and this, along with food and air would be shipped out to the asteroids, where no manufacturing or agriculture is likely to take place. On the third leg, the asteroids would send raw materials back to Earth (it is unlikely Mars would be a customer, as it has plenty of its own resources).

DIGGING IN SPACE

Something that is discussed a lot less than the difficulties of getting to and from a space-mining site, whether it is on the Moon or the asteroids, are the practical issues involved in actually extracting the materials. Most plans become a little handwaving at this stage, and yet asteroid miners face a very different environment from that experienced in a nineteenth-century gold mine where all that was necessary was a pick, a lamp, and some muscle power. On the asteroids, the negligible gravitational field makes every physical undertaking more difficult—blasting, for instance, would result in chunks of asteroid flying off in all directions without sufficient gravity to keep them in place. And anything that has a Newton's third law push back—like using a pick or jackhammer—is likely to send a miner floating off into space unless they are well secured.

There is also the nature of the material that is being worked on an asteroid, which can differ considerably from the familiar rocks that miners have handled on Earth. Our experience with lunar soil or "regolith" (the term is derived for the Greek for "rug" or "blanket" as it refers to the thin layer of material on top of the bedrock) on the Apollo missions is that continual bombardment by meteorites has left behind a collection of sharp glass-shard particles that are very abrasive, making regolith a danger for any machinery with moving parts. (It is the similar glass shards in volcanic ash that give airlines such problems, as they tend to be sucked into the engines, partly melt, and cause moving parts to seize up.)

Astronauts on the Moon were also surprised to discover that the lunar soil stuck to them—and pretty well everything else—whenever it was disturbed. This is because the radiation bombarding the surface tends to give the regolith an electrical charge, so the particles sticks to objects in the same way that pieces of paper are attracted to a charged balloon. This is also likely to be true of any surface dust on asteroids.

It might seem that the obvious solution to dealing with regolith is to use something like an industrial vacuum cleaner to suck up the material to be mined, pulling it into a collecting bin. Unfortunately, vacuum cleaners don't work without an atmosphere. They don't simply suck, but rather reduce pressure inside so that the atmospheric pressure on the outside pushes the material in. With no air to produce that pressure there is no suction. An alternative approach has been proposed, using a dual tube that feeds gas to the surface then pulls it back, carrying dust with it—but this is the kind of challenge that will be difficult to test for real without getting the equipment out into space. The sometimes-derided ISS could find a new role as a handy local test bed for asteroid mining technology.

EXPANDING THE FRONTIER

If we discount Venus as an impractical destination for miners as much as for colonization, we can think of the Moon, Mars, and the asteroid belt as representing the equivalent of opening up unexplored parts of our own countries. This was the kind of thing that was still possible in the United States in the nineteenth century. But taking in a wider swath of the Earth—going from Europe to Australia, for instance—involved significantly greater resources, and the same goes in the solar system for the outer planets.

After the belt we encounter the four gas giants: Jupiter, Saturn, Uranus, and Neptune. One thing is very clear. We are highly unlikely ever to establish a colony on the planets themselves, even though they could provide rich pickings for specialist mining missions. Not only are the conditions too harsh, these planets are not called gas giants for nothing: They are largely insubstantial with no solid surface in the conventional sense. Take Jupiter, for instance. The closest of the gas giants, it is also the largest planet in our solar system, with a mass that is more than twice that of every other planet combined.

Like a small cold star, a lot of Jupiter's mass is made up of hydrogen. It may have a rocky core, but if so it is thought to be relatively tiny, with the rest of it made up of liquid and gas. With no certain true surface, it's hard to say what the surface pressure is, but it is likely to reach as high as ten thousand times that of the Earth's surface by the time any core is reached. In terms of exploration, we should never say "never," of course. Technological developments can often outstrip our imagination. In a 1950s science fiction novel, the author James Blish makes the casual remark, "there was no electronic device anywhere on the Bridge since it was impossible to maintain a vacuum on Jupiter." The assumption Blish made was that electronics implied vacuum tubes, which would collapse under the atmospheric pressure. A few years later and solid state electronics would make this particular "never" outdated. But even so, Jupiter and its fellow gas giants show no signs of being able to provide habitation.

There is a possibility, though, of building bases on the outer planets' moons, at least to act as temporary shelters for miners. Titan, the largest moon of Saturn, is the size of a small planet (it is bigger than Mercury), has a thick, mostly nitrogen atmosphere and is well provided with the resources necessary for life. For that matter, Europa, one of Jupiter's large moons, has a surface made of water ice and it has been suggested that the tidal energy provided by close proximity to the massive surface of Jupiter means that there is a liquid water ocean underneath the ice, despite the distance from the Sun.

Europa is probably the most likely habitat for primitive life to exist outside of the Earth in the solar system—but it is not a great place for humans to colonize. Not only is there a lack of accessible resources other than water, it suffers from the same problem as most of Jupiter's moons—it is constantly bombarded with radiation. The combination of cosmic rays and Jupiter's own significant radiation belts mean that living there without vast amounts of shielding would be impractical.

THE JOYS OF HELIUM 3

While there are many opportunities for exploration and discovery around the outer planets—Europa is certainly somewhere we want to find out more about, for instance—they have very limited appeal as part of the practical frontier. It would be possible, but difficult, to "mine" some gaseous resource, notably the helium isotope helium 3, which is rare on Earth but likely to be much more abundant in the gas giants. As we have seen, this is potentially very useful for nuclear fusion, the power source of the Sun and in the long term likely to be the Earth's main source of energy.

It has been estimated that there is enough helium 3 in Jupiter's atmosphere to produce around 5 billion terawatt years—enough energy to satisfy Earth's requirements at current levels for many millions of years. (This assumes, of course, that we can develop effective nuclear fusion power stations to make use of the fuel, which should be practical in the sort of timescales in which we could consider "mining" helium 3 from Jupiter.) Human beings couldn't get too close to Jupiter because of the radiation levels, but unmanned ships, especially if designed to cope with the extreme atmospheric conditions on Jupiter might be able to satisfy Earth's demand for power into the indefinite future.

Although current experiments in fusion reactions tend to use hydrogen isotopes, these have the disadvantage of producing dangerous high-energy neutrons, which can't be easily steered (because they have no electrical charge) and which will soon make the reactor radioactive. However a helium 3 fusion reaction only has a proton (a hydrogen nucleus) as a by-product. Because they are charged, protons can be controlled and dealt with easily. The downside of helium 3 fusion is that it requires much higher temperatures than currently contemplated fusion methods, but it may be considered a valuable fuel in the future.

Some of those who advocate space colonies see helium 3 collection as providing a major economic benefit, but the high reaction temperatures mean it has its own problems as a nuclear fusion fuel, and it may not be worth the expense of recovering it from a gas giant. It would certainly require energy prices on Earth to soar far above current levels. Without a market for helium 3 there would be no commercial benefit in establishing mining bases in the outer planets because they are too distant and hostile.

The sheer distance and energy required takes the outer planets far beyond a goal like Mars. With current technology, Jupiter is over 2.5 years away, and Saturn around 6 years' travel time. Even with the potential advances in propulsion technology that could be envisaged within 100 years, Saturn is likely to take at least 2 years to reach. The gas giants and their moons would be of huge scientific interest, but would not be suitable for colonies or as a resource supply. At best they could be considered way stations on the potentially much more fruitful, but vastly more difficult, exploration beyond the solar system.

AN UNCLEAR PATH

It is easy to get carried away with the possibilities of commercial ventures to Mars and the asteroids and forget that NASA, and its competitors from Russia, Europe, Japan, and China are still active. Although governments might be less likely to send out mining missions, they could provide the infrastructure—on, say, Mars—that would make it a much more practical possibility. One of the problems for NASA is that its pathway to the stars was flawed by bureaucratic bloat and sabotaged by a lack of political will. Politicians were happy to talk about the potential to return to the Moon and go on to Mars, but stumping up the money required was a contentious issue.

When the Space Shuttle program was closed down, the intention was to replace it with a program called Constellation. In fact this was much more than a replacement. The space shuttle was, in retrospect, a blind alley, a craft that was only ever intended for playing in the Earth's backyard. It was necessary to come up with a replacement that could put the Moon and Mars sensibly in

NASA's sights, and Constellation was intended to be that vehicle. Its main components were to be the Ares I and Ares V boosters, the modern equivalent of Saturn V, and Orion, a manned capsule for the twenty-first century.

In effect, Orion was a larger version of Apollo's command and service modules—the bits that stayed in orbit, while a further vehicle, Altair, was planned as a lander to reach the Moon's surface. Since 2010 the Constellation Program has been at best on hold. Instead the intention is to rely significantly more on commercial lifters like SpaceX's Falcon 9, while designing a replacement for the Ares concept called Space Launch System, based on shuttle launch technology, providing a range of boosters that could be utilized by a variant of Orion known as the Orion Multi-Purpose Crew Vehicle. Whether this new vision will survive or be replaced yet again as politicians change their minds is so far unclear. It emphasizes the incompatibility of these large-scale, long-term expensive projects with the short-term thinking generally employed by the political class.

Not every journey of exploration—even to the stars—needs to involve a vast budget, though, as our early and so far most far-reaching probes like *Voyager 1* have shown.

9.

PROBING THE GALAXY

||||||||||||||||||||||||||||||||||||||||||||||||||||||

Damn the Solar System. Bad light; planets too distant; pestered with comets; feeble contrivance; could make a better myself.
> —Attributed to Lord Francis Jeffrey (1773–1850)

At the moment, *Voyager 1,* our most distant probe, is around 10 billion miles from Earth, heading out at 38,000 miles per hour. At this speed it would take around 73,000 years to reach the nearest star, but it will lose its ability to power up its instruments in about 10 years' time. *Voyager* was never intended to enable us to explore the galaxy, but the first generation of spacecraft intended to take a human presence truly beyond the solar system are likely to be robotic probes for which *Voyager* was the outstanding forerunner. If we ever want to reach beyond the solar system we need to get our probes—and even manned ships—heading out faster and communicating for longer.

A DIFFERENT SCALE

It's a big challenge—in fact "big" is the operative word throughout the next two chapters, which look at different aspects of reaching for the stars. The universe is massive on a scale that can easily be overwhelming. As we have seen, Archimedes reckoned the universe was around 1,800 million kilometers (1,118 million miles) across, comfortably incorporating Jupiter, but not extending as far as Saturn. A guess it was, though, because in Archimedes' day there was no appropriate measuring stick to reach out to the stars. That hardly seems surprising. How is it possible to look out at a set of pinpoints of light and decide how far away they are? The first attempt came down to using the way things appear to move.

By the start of the eighteenth century they were starting to use parallax to measure the distance of the nearer stars. This is the effect you see if you hold a finger up at arm's length and look at it with each of your eyes in turn, closing the other eye. The finger seems to move against the background, and you can use the distance between your eyes and the angle the finger appears to move through (plus a spot of geometry) to work out how far away the finger is. The same can be done with stars, but rather than relying on the distance between a pair of eyes, the parallax is generated by the distance between the Earth's position on either side of its orbit—a distance of around 300 million kilometers (186 million miles). Using this technique we can work out that the nearest star is around 40,169,669,040,000 kilometers (25,000,000,000,000 miles) away.

As this is an impractical kind of distance to work with—and we are talking about the nearest star here (other than the Sun)—astronomers tend to measure distances in light-years, the unit I have been casually using for a while without mentioning any detail. A light-year is just the distance that light travels in a year. This makes that next nearest star after the Sun, Proxima Centauri, a more manageable 4.243 light-years distant. (To make matters even more confusing, astronomers often tend to use parsecs to describe distance, a measure that is directly derived from the angle of parallax of an object when viewed from the Earth. Proxima Centauri is 1.3

parsecs from us. But for most normal purposes the light-year is more accessible and I intend to stick with that.)

The only slight danger of representing the distance in light-years is that it makes a relatively near star sound like a distance that is instantly achievable. But bear in mind that the light, the fastest thing in normal space, still takes over four years to make the journey from Proxima Centauri. A useful scale can be to compare the distance to Mars—seen as quite a challenge at the moment—with the distance to Proxima Centauri. If the distance between Earth and Mars, when they are at their closest in 2018, was represented on a map as 1 meter (3.28 feet), then Proxima Centauri would be 718 kilometers away requiring 446 miles on the map to represent its distance. In orders of magnitude, when we are thinking of local star travel, rather than pottering around inside the solar system, we need to increase journey lengths by a factor of at least 10,000.

Parallax can only work so far, to give distances to relatively close stars. Beyond that, astronomers use the rather odd-sounding "standard candles." The concept is straightforward—if I have two identical candles and one is closer than the other, I can use their relative brightness to work out their distances. The same goes for identical stars. If I know the distance of a closer star from parallax, I can work out how far a more distant star is by how much dimmer it looks. The trouble, though, is how to discover if two stars have the same actual brightness. We can hardly go out there and check them.

What was necessary to use standard candles was to discover stars that have an identifying characteristic that is linked to their brightness—and the first that fitted the bill were variable stars. These are stars that get brighter and dimmer at a steady rate, and it was discovered that the speed of variation of some variable stars, notably the Cepheid variables, named after the constellation Cepheus, had a good match to their brightness. So by finding two Cepheids with the same rate of flashing, it is possible to use their relative brightness to distinguish distances far greater than parallax can measure. Different types of standard candles provide most of our interstellar distance metrics.

The scale of our galaxy, the Milky Way, makes the distance to Proxima Centauri seem pretty much like a casual afternoon stroll. The Milky Way is a spiral disk, around 100,000 light-years across and around 1,000 light-years thick, containing as many as 400 billion stars. That is quite a region to explore. And yet if we ever truly decided to explore the universe, rather than just our own galaxy, then the Milky Way too is nothing more than our close environs. As we have seen, the nearest neighboring galaxy of comparable size, Andromeda, is around 2.5 million light-years away. Go further still and there are billions of galaxies in the universe, each with between tens of millions and trillions of stars. Or rather, there are billions of galaxies in the observable universe—because the universe is so big, we can't tell just how big it is.

The problem is that light travels at a finite speed. We think the universe has been around for 13.7 billion years, so if you look in opposite directions out into the universe, the furthest light you could ever see is the photons that have been coming toward us for 13.7 billion years. At first sight this might seem to make the visible universe 27.4 billion light-years across. (This is why light-years are much more useful than parsecs, unless you are an astronomer, or just ornery.) But there is a catch. The universe is expanding, and has been doing so for all of its existence. Some of that expansion is thought to have been faster than the speed of light. (Nothing can move faster than light through space, but this limit, emerging from Einstein's special relativity, does not apply to the expansion of space itself.) And the result is a visible universe that is significantly bigger than the time available suggests.

If light had been coming toward us for 13.7 billion years it would actually have traversed 46 billion light-years, making the "visible universe" 92 billion light-years across. How big the universe actually is remains a mystery—it could be infinite in scale, and some cosmologists believe this to be the case, but equally it could be finite but unbounded. This is a slightly mind-bending concept, but the universe is being effectively bent back on itself in an extra dimension. A good model is the surface of the Earth, which is also finite and unbounded (just in one fewer dimension). It doesn't have any boundaries—you will never reach the edge however far you travel—but it is finite in scale. That is a two-dimensional surface, twisted through a third dimension, where such a universe would have three spatial dimensions, warped through a fourth.

ONE GALAXY AT A TIME

With a good picture of the scale of the challenge facing any interstellar exploration we can see that realistically, without some distance-busting hyperdrive that makes it possible to cover millions of light-years in a few years of personal elapsed time, we are limited to the Milky Way as a frontier to explore. The scale here is still one that it is easy to struggle to get your mind around. If 1 in 3 stars in the Milky Way has some form of solar system, that gives us around 100 billion stars to take a look at. To put that in context, if an exploratory mission took in a new star every day, it would take over 273 million years to visit them all. Space is not just the final frontier, it is the ultimate frontier. It is not one that is ever realistically going to be exhausted.

Even if we stick to our stellar near neighborhood—say a radius of 30 light-years, which brings hundreds of stars (though not all with planets) into reach—a ship would need to reach sizable speeds to achieve the journey in anything like an acceptable time. *Voyager 1* would take around 75,000 years to reach Alpha Centauri (though it is not in fact heading there). We need to go much faster than this. But to do so implies packing a huge amount of energy into our means of propulsion.

To get a feel for this, let's say we wanted to send a ship across our 30 light-year neighborhood. Imagine that we have a new wonder engine that allows us to achieve a phenomenally fast speed— half the speed of light, 150,000 kilometers per second (93,000 miles per second). Then the ship would still take 60 years to reach its destination, at least, as far as the people left behind on Earth were concerned. The good news for those on board (assuming this is a manned probe) is that relativistic effects are starting to kick in at these speeds. Einstein's special relativity tells us that clocks slow down when you move quickly. The crew would only think that 52 years had gone by. If we imagined an even more extreme engine that managed to get the ship up to 90 percent of light speed (0.9 c), the journey would take only 33 years as seen from Earth, and an impressively quick 26 years for the crew. But the energy cost involved would be immense.

To take that extreme example of 0.9 c, assuming we had a 100-ton ship (about the same weight as the loaded space shuttle used to be), it would take 12 billion trillion joules of energy to get up to that speed. This number is large enough to be meaningless, but to put it into context is the equivalent of the output of every power station currently operating on Earth running for 830 years. This is the sort of challenge that spaceship designers have to face if we are to get a probe to the stars in any reasonable timescale. Of course the requirement drops with lower speeds and distances—but even so, the challenge is immense.

BEATING TECHNOLOGY LAG

A big problem that looms for anyone hoping to develop the equipment to get probes to the stars is that our spaceships have traditionally suffered from massive technology lag. This always has the potential to be a problem with projects with long lead times. When

I worked at British Airways there were two clear examples of this. The first was Concorde. When I used to take friends around the BA engineering base in the 1980s they all wanted to see Concorde, something very special as the only supersonic airliner in operation. I was lucky enough to be able to take them on the flight deck—and the reaction was one of shock every time. Because despite being the most advanced airliner ever built, Concorde's instrumentation looked like something out of a World War II aircraft. It was purely mechanical. No screens at all. The design was much older than the times in which the aircraft operated.

We had the same problem with onboard computing technology. When the airline first decided to put PCs on board its planes in conjunction with aircraft manufacturer Boeing, it specified the latest hot technology. But by the time the planes were finished, the computers that were in the design specification were around ten years old. Imagine buying a ten-year-old PC today. If your designs are fixed many years in advance you will inevitably be left behind by the rapid advances in technology, particularly information technology, that arrive all the time.

Things were even worse when it came to space technology, as a similar computing problem demonstrates. Until 1997, space shuttles used computers with magnetic core memories, an antiquated solution to storing data that was common in computer rooms in the 1960s but only turned up in museums by the 1990s. As long as space travel remains a rare, high-expense event, developments are likely to involve long lead times and to be in danger of incorporating out-of-date technology—especially if they are being built by a slow-moving, large organization like NASA. To get around it, the two main requirements are to make space vehicles cheaper, with a quicker design-to-build time, and to ensure that any technology that does date horribly quickly like computers can be slotted in at the last minute, rather than designed into a system ten years earlier.

SEARCHING FOR A NEW PARADIGM

There is probably another worrying parallel to be found in the aviation industry. Concorde provided an unprecedented leap forward in the ability of ordinary people to travel at speed. The maximum it was possible to fly at went up from around 600 miles per hour to 1,400. This happened over around twenty years. If that kind of pace of advance had continued, you might expect some commercial airline passengers to be flying at around 5,000 miles per hour today. Instead we have dropped back to the kind of speed that was already possible in the 1960s. (Remember, the Boeing 747, still widely in use, is a 1960s design.) This is because airline technology has not gone through any radical changes in this period, apart from that one great supersonic leap forward that was stymied by political pressure rather than engineering challenges.

If you now think of the way we travel through space, we have not really advanced from sending our astronauts riding into the sky on converted ballistic missiles. The launchers may be more sophisticated, but they have really hardly advanced. Most missions currently use very similar technology. And though there are perhaps the equivalents of Concorde coming through with the likes of the SABRE engine (see page 56), to make the real leap forward necessary to reach the stars we need to see a significant change in the technology we use for traveling into space.

In general, unless we are prepared to consider highly speculative modes of travel where the speed of light is no longer a true barrier, any rocket-powered probe, whether or not it attempts to leave the solar system, is limited by a simple equation that describes the effectiveness of a rocket motor. The detailed mathematics is not important (and rather dull)—the crucial aspect is how much velocity can you get out of your mass ratio, which is simply the ratio of "wet mass" (total mass at the beginning of the voyage including fuel) to "dry mass" (the actual payload).

A high-mass ratio is good in the sense that there is the maximum amount of available fuel to provide power, but it limits the ship's cargo capacity. And, of course, a crucial component is how much extractable energy there is in any particular mass of fuel. The more energy, the less fuel required to get to any particular velocity. It is also possible to "cheat" by looking for ways to increase speed that don't use fuel (like solar sails) or to avoid carrying the fuel for the whole journey—by picking it up from dumps along the route, or by scooping it up from space—but usually in an interstellar voyage these are likely to be add-ons and boosts, rather than the main contributor to the propulsion.

PACKING IN THE ENERGY

It is important, then to know just how much energy there is in a fuel. The numbers can be surprising, because substances we think of as packing a punch aren't necessarily the best sources of energy. If you were, for instance to compare the explosive TNT with gasoline, neither, I admit, intensely likely to be used to fuel a space vessel, it might seem obvious that the TNT produces a much more dramatic outflow of energy. Yet gas actually has 15 times the energy per unit weight of TNT. The only reason TNT makes a better explosive is that it releases the energy it has much more quickly.

Of the typical fuels that are or may be used in a spaceship, hydrogen—the most potent in current rockets—has around 2.6 times the energy for a particular weight than gasoline does. (Though hydrogen is not without its problems, as even as a liquid it takes up 4 to 5 times the volume of a hydrocarbon fuel like gas. This isn't significant in a ship that spends its life in space, but it does matter when it comes to taking off from Earth as more volume means a bigger metal container—think of the huge fuel tanks that were used on a shuttle launch—and that means a lot more mass to carry around.)

We also have to bear in mind that anything that burns in a traditional sense, be it gas or hydrogen, needs oxygen to complete the combustion equation. Burning is really just a form of high-speed oxidation—a substance reacting with oxygen. The need to carry oxygen is a nontrivial addition as it is significantly heavier than hydrogen, so carrying oxygen adds hugely to the mass burden, which is why the proposed Skylon rocket plane (see page 56) takes as much of its oxygen as it can from the atmosphere before the gas pressure becomes too low.

A lot of hope has to be placed on the possibility of powering the next generation of space vessels with nuclear power. It wouldn't be too much of an exaggeration to say that taking this step is the only way to get past our tie to traditional rocketry. Just as it has been realized that the only way to keep submarines at sea for months at a time is with a nuclear power plant, so too a long-range spaceship is likely to need nuclear power. There is a good reason for this. Straightforward nuclear fission, as used in current nuclear power stations, produces around 2 million times as much energy from the same mass of fuel as does gas, while the more advanced nuclear fusion that powers the Sun (and is the proposed power source of the Icarus starship—see page 228) produces 6 million times the energy. And neither of them require an oxidant.

Even the possibilities of nuclear power, though, pales into insignificance when put alongside the energy released when matter and antimatter combine. Here we are talking of a 2-billion-fold increase in energy output over gasoline for the same mass of fuel. The main problem with antimatter is simply obtaining enough of this rare substance to do anything practical with it, as the current world production is measured in millionths of a gram a year.

A second challenge that applies when dealing with both antimatter and nuclear fuels is safe storage. Antimatter has to be kept away from normal matter to avoid instant annihilation, while fuels for nuclear fission present their own familiar storage problems due to their radioactivity. By far the biggest consideration, though, with all these fuels offering a dramatic increase in energy output per unit mass of fuel is how to build a motor that will convert the indubitably dramatic amounts of energy they give off into a means of propulsion.

BLOWING YOURSELF ACROSS THE UNIVERSE

A technology that has been hovering in the wings, waiting to be usable almost as long as nuclear fusion power plants have been under development, is the idea of powering a ship—and potentially a starship—with a series of small nuclear explosions, a technique usually called pulsed nuclear propulsion. One of the reasons what to some seems absolute madness was first considered was a rather desperate attempt by those who almost fanatically supported fusion bombs—hydrogen bombs—to save their precious technology as the population turned against the horrors of such devastating sources of mass destruction. Instead the scientists tried to find "friendly" applications of the H-bomb, from mining to powering spaceships.

There is no doubt that nuclear explosions produce plenty of power—the question is how to harness that power to drive a ship, rather than to blow it up. As early as 1947, Stanislaw Ulam and Frederick Reines, working at Los Alamos, New Mexico, produced a concept by which a ship might be driven by a series of small nuclear blasts. The idea is that the back part of the ship would contain a vast stack of small nuclear devices, which would be pushed out of the rear of the ship at regular intervals. Before being triggered, each device in turn would be shot into position behind a large, curved plate, a so-called pusher plate, fixed to the rear of the ship.

The pusher plate was a (usually) curved sheet of steel, coated in something like graphite. When the bomb was tens of meters behind the ship it would be exploded. Debris and vapor would blast toward the plate from the explosion, transferring momentum to the ship as it bounced back off. The result would be that each explosion would give the ship a new push, increasing its velocity. Some sort of shock absorber between the plate and the ship would ensure that the passengers were not liquefied by the sudden and intense acceleration from the explosion. This only left the slight problems of ensuring that the astronauts were not given a deadly blast of radiation, and that the explosion wasn't a touch too large, or too close, vaporizing the pusher plate or, even worse, blasting apart the storage for the other nuclear devices and setting off a massive chain reaction.

In those early heydays of nuclear bombs there was even a chance to test out this concept for real. A "small" bomb, around 20 kilotons, similar in power to that used on Hiroshima, was exploded on the South Pacific island of Eniwetok, home of the U.S. bomb tests once they moved from New Mexico. Two steel spheres, coated in graphite, were hung from wires 10 meters (32.8 feet) from the bomb. Although the wires were totally vaporized, the spheres rode the blast wave and were found with undamaged interiors several kilometers away.

After the initial idea, the concept was developed in a number of directions. One simple improvement, devised in 1955, was to add a powerful magnetic field to the pusher plate. As the vaporized output from the bomb would mostly be charged ionized particles, the magnetic field would be used to reduce wear and tear on the pusher plate, as most of the "push" would be transferred through the magnetic field without the material ever coming into contact with the plate. Others, though, felt that there was something wasteful and frankly amateurish about throwing the explosive devices out of the back of the spaceship and wasting most of the blast in other directions.

The alternative was to go for an explosion within some kind of combustion chamber. Apart from any consideration of wasting energy, because of existing technology, internal combustion seems the more sophisticated approach—compare, for instance, a steam engine with an automobile. But there is obviously a tiny issue. Setting off a nuclear device in open space at some distance behind your craft is one thing—setting it off inside the spaceship is a far more worrying concept (especially with many more bombs stored in the same engine room). On the other hand, if this were possible, then far more of the energy released in the explosion could be employed in driving the ship.

Most designs involved injecting a propellant like water or hydrogen into a powerfully shielded chamber. The bomb would be exploded in the center of the chamber, vaporizing the propellant, which would both provide the reaction material thrown out of the back of the spaceship and help protect the chamber itself from being blasted apart. While in theory feasible, the practicalities of making such a motor survivable, both in terms of the explosion and the associated radiation are complex, and there is also the need to carry large quantities of the reaction propellant, which would significantly add to the mass and size of the ship.

THE MIGHTY ORION

It is not surprising, then, that the first detailed design based on this principle relied on the earlier idea of using external explosions. Despite being a proof of concept project undertaken as far back as 1958 to 1965, this is still frequently brought up as the exemplar of nuclear pulse designs—its name was Project Orion (no relation to the present NASA Orion capsule design). This was a dramatic response from the United States to the initial shock at the Soviets getting a satellite into space first before anyone else. Just how far-reaching the Orion vision was back then can be seen from the project motto: "Mars by 1965, Saturn by 1970." It makes even the now apparently highly exaggerated futurology of *2001: A Space Odyssey* seem tame.

The original design for Orion was vast by the standards of anything that has ever been built. The space shuttles weighed around 100 tons each—Orion would have weighed in at about 10,000 tons. It was designed to carry up to 150 people with a payload capacity of thousands of tons. This was the kind of ship that would make setting up a colony on Mars (provided you could work out how to land there) a trivial exercise. This was a *real* spaceship.

Unthinkably from a modern viewpoint, despite the requirement to fling out a string of nuclear bombs behind its 40-meter (130 foot) pusher plate, the original plan was that Orion would take off from the Earth's surface, making use of a nuclear test site to lift off through the atmosphere on its stream of explosions. The fission devices used to propel it would start with 0.1 kiloton devices every second at take-off, building up to 20 kilotons every 10 seconds. Six times a minute it would set off devices more powerful than the Hiroshima bomb.

The original design for Orion involved carrying 2,000 nuclear pulse bombs, and it was expected to come in at less than the $20 billion that the Apollo program eventually cost. This was imagination soaring to the heights—not so much paper NASA as fairy-tale NASA—but politics and funding soon brought the concept crashing to Earth. Initially it proved difficult to get enough funding to make a serious start on the project and it was handed from pillar to post. What had originally started in the skunk works at the private company General Dynamics and had briefly ended up with the Department of Defense's Advanced Research Projects Agency (ARPA, the forerunner of the current DARPA), and then the U.S. Air Force, was finally handed over to NASA in 1963.

SHRINKING TO FIT

The five-year-old agency had its own priorities and approach at this point, so to fit with the NASA perspective, the Orion team proposed a new, much smaller implementation that could be blasted into space on the Saturn V being developed for the Apollo program. This took away the scary concept of taking off from the surface on a stream of nuclear explosions, but reduced the mass of the vehicle to around 100 tons and used a much smaller pusher plate to fit within the diameter of the Saturn launcher, meaning that there would be much less power available. Even with the reduction in scale there was too much of the new Orion to carry the craft on a single Saturn, so the idea was to split Orion into two or three parts and assemble them in space. One possible mission that was scoped out was to take the Orion to Mars with eight astronauts. Even the cut-back version would vastly improve on the abilities of today's rockets. Where our twenty-first century Mars missions are expected to take 6 to 9 months one way, the 1960s Orion would have executed a round trip in 125 days.

While the move away from blasting away from the Earth using a string of atomic bombs was certainly an improvement from both PR and health and safety viewpoints, cutting back on the scale of the original Orion to such an extent was probably a mistake, and any future reboot of a similar concept would surely be better carrying many more sections into space. Even if we were not able to replicate the massive scale of the first Orion designs, it is arguable that any ship making use of the vast power available with a nuclear-pulsed propulsion should take the opportunity to carry a truly sizable payload.

There still was a worrying "eggs in one basket" aspect to the NASA compact version of Orion. The propulsion unit, holding hundreds of atomic bombs, was supposed to be sent up on board a single Saturn V. If something went wrong with the launch or in the early part of the flight, particularly if the rocket exploded as several had in those early days of Saturn testing, this seemed a horrendous risk, and one that will still need to be addressed at some length if this technology is ever to be used in practice. And that is still, as we shall see, a distinct possibility.

Something that wasn't envisaged in the days of Orion was the move toward using a wider range of smaller, cheaper launch vehicles rather than the single, massive Saturn V. This means that it is more practical now to envisage transporting small numbers of nuclear pulse propellants (okay, atomic bombs, but NPPs is more media-friendly) at a time, rather than the large numbers required for a lengthy mission. The devices could be transported incomplete, so that even if the ship carrying them crashed there would be no danger of explosion. All that would then remain is to prevent any radiation leaks, something that is highly practical, at least for collisions. In the event of an airborne explosion, there would be some Chernobyl-style dispersal of radioactive material, which suggests that care needs to be taken in the selection of both takeoff location (ideally somewhere very remote) and prevailing weather. But the risks should not be much greater than carrying nuclear devices on bomber aircraft, as has been undertaken since the 1940s.

The alternative, should appropriate fissionable materials be available on the Moon, an asteroid, or Mars is not to take the bomb-making kit up from Earth at all, just the tools to perform precision engineering, and to manufacture the final device off Earth, so that there is no danger of a crash or explosion spreading contamination. We know that there is uranium on the Moon, detected by the Japanese *Kaguya* spacecraft in 2009. This was something of a surprise as the Moon is significantly less dense than the Earth, so the expectation was that there wouldn't be much in the way of high-density elements like uranium. However there is likely to be much more radioactive material available on Mars and in the asteroids—the downside being that we have to get there first before being able to start on construction of our high-power spaceship.

In practice, although safety probably could have been addressed in a fashion that would have satisfied the looser restrictions of the 1960s, another obstacle was about to be put in Orion's way just as NASA took the project over, one that would sink the project for decades to come. The Nuclear Test-Ban Treaty of 1963 made it impossible to run trials of the Orion power system—essential with such a novel approach—and though in principle an argument could have been made for Orion being an exception if it could be shown not to have military applications, it was a blow that put political opinion firmly against the project. Realistically too, just one year after the Cuban Missile Crisis, with suspicion about nuclear arms still paramount, it would have been hard to convince anyone, let alone the Soviet Union, that Orion was anything other than a sneaky U.S. attempt to construct a massive, multi-warhead nuclear missile.

The difficult sell of the nuclear bombs, the risks of getting them into space, and the restrictive treaty combined with the fact that because of its military origins (Orion was, at the time, classified)—not to mention NASA's increasing focus on the Apollo program—mean that, with hindsight, Orion was always doomed. And yet it is hard not to admire the sheer chutzpah and imagination of the designers of this project, especially in its original colossal form. This was pioneering with a passion.

SURFING ON THE POWER OF THE SUN

The explosives that would have been used to produce the thrust for Orion were nuclear fission bombs, however this was only ever seen by the nuclear-pulsed-propulsion enthusiasts as a stepping stone to using nuclear fusion, the power source of the Sun. The great advantage of using fusion explosions is that where a fission reaction relies on having a critical mass, establishing a minimum size below which it is impossible to shrink the explosive, fusion has no lower limit. Provided enough temperature and pressure is applied, any size of reaction material can be forced to undergo fusion and expel large quantities of energy.

A ship that was powered by fusion explosions would not need such a huge pusher plate or such extreme defenses against the explosion as one relying on fission. A less colossal, more practical design could ensue when the explosions were as little as a thousandth of the power of those required by a fission pulse engine. This was the thinking behind a 1970s proposal building on the Orion concept called Daedalus. At a stroke, by moving to fusion one of the biggest problems facing Orion was removed from the equation. There was no concern about the dangers of transporting dangerous fissionable material into space, with all the risk involved should there be an explosion or a crash. The fuel used in a pulsed fusion motor would be relatively harmless isotopes of hydrogen and helium.

The main aim at the time was to design a ship to travel one way to Barnard's Star, which at around 6 light-years away was thought in the 1970s to be the nearest star likely to have a solar system (it has since proved not to have one). The unmanned probe, devised by the British Interplanetary Society between 1973 and 1978, was planned to take 50 years for the journey, powered by a fusion reactor. This journey time implied getting up to around 12 percent of the speed of light. The propulsion was designed to work by igniting pellets of a mix of deuterium and helium 3 (hence the interest in helium 3 "mining" discussed on page 120). These would be forced to fuse using a process known as inertial confinement fusion, one of the two main approaches still taken in fusion energy research today.

Typically attempts to use this technology to produce a power station heat and compress the target material using extremely powerful lasers. The lasers zap the outer layer of a pellet, which explodes against the rest of the pellet, compressing it and creating powerful shock waves that, if intense enough, produce nuclear fusion. The most impressive of these experiments at the time of writing is at the National Ignition Facility at the Lawrence Livermore laboratory in California. But as yet these devices have not produced enough energy to be self-sustaining—they require more energy to get the fusion to work than they give out. The other problem with translating this technology to spaceship propulsion is that at the moment the lasers and support equipment required are huge, taking up whole buildings. The Daedalus design envisaged using electron beams rather than lasers, in part because at the time laser technology was largely unknown, though this now seems a less practical approach.

The difficulties at the National Ignition Facility underline the biggest drawback of fusion-based nuclear pulse propulsion. While fusion does away with many of the safety and PR issues that dogged Orion, it has one huge issue of its own. Yes, the ship could be a much more practical size, the whole setup would be easier to test and crucially the components of the ship, or the whole vessel could be put into orbit without risk. But the fact is that even back in 1958 we knew how to make the explosives that would power Orion. Given the money and the political will it could have been produced there and then. The problem faced by Daedalus and its successors is that fusion technology is still beyond our grasp, even on the scale of a massive earthbound laboratory, let alone in a compact motor that could be fitted to a spacecraft. It is entirely possible that we are still fifty to a hundred years or more away from being able to build such an engine.

Assuming such problems could be overcome, Daedalus would appear much more like a traditional engine-powered ship to an outside observer. Rather than pushing out sizable bombs, the pulsed propulsion unit was planned to work its way through 50,000 tons of pellets, fusing 250 a second in a storm of miniatures explosions that would seem like a continuous stream of a rocket to the human eye. But that is a very big assumption, and just as for the moment we need to continue with nuclear fission for our power plants on Earth, because fusion is simply not manageable, so the even bigger challenges of producing fusion in a compact, light spaceship engine mean that we may have to revisit those rather more scary fission concepts first if we are to get an effective starship underway in a realistic timescale.

TO CARRY FUEL TAKES MORE FUEL

Even with nuclear pulse propulsion, one of the biggest problems faced by starship designers is that you end up spending most of your energy getting your fuel to move. In the 1970s Daedalus starship design, only 4,000 out of a total of 54,000 tons that had to initially be accelerated was the actual ship and its payload. All the rest was fuel. And much of that 4,000 tons was the engines, leaving a mere 450 tons for the payload. There's a terrible circular conundrum—the further you want to go, the more fuel you need to carry, the heavier the ship gets, the more fuel it needs to carry.

The obvious solution is to minimize the amount of fuel on board. This is the idea behind the SABRE engine (see page 56), which at least saves on having to carry so much oxidant to make the fuel burn. One possibility is to pick up fuel as you go. We have already looked at the possibility of providing fuel dumps from asteroid mining, but this would not be hugely useful well into an interstellar flight. However, space is not empty, but has a fair amount of scattered material even between the stars, much of it hydrogen. A 1960s concept with a pure space opera name, the Bussard ramjet, was designed to make use of this.

The idea was to get a ship up to high speed using conventional fuel, then to use the movement of the ship through space to scoop up and compress hydrogen to such an extent that a fusion reaction begins. This way an approach similar to the fusion pulsed engine is achieved but with a more continuous stream of fuel through the engine. The good news is that if such a mechanism could be made to work, it could achieve a steady acceleration of as much as 1 g for as long as is required. This has two excellent implications. If you can accelerate at a constant 1 g there is no need to mess around with complex rotations to provide artificial gravity—the acceleration does it for you and keeps your occupants healthy.

What is more, although 1 g doesn't sound like much acceleration, it soon adds up to build an impressive speed. Such a ship could reach Alpha Centauri (as long as it didn't bother to slow down) in three years from the astronauts' viewpoint. If the acceleration (and the hydrogen) continued indefinitely, the ship would reach the Andromeda Galaxy in just twenty-five years of the crew's time (though because of relativistic effects, it would seem much longer—millions of years—to those left back on Earth).

It all sounds great, but as always with this kind of imaginative design, making it work is very different from the basic concept. There probably isn't a high enough density of hydrogen in space to feed the ship as it went along, and even if it was just about possible, the scoop would have to be huge, extending many kilometers out from the front of the ship. Typical designs ranged in aperture from 150 to 4,000 kilometers (90 to 2,500 miles) wide. This would clearly be too clumsy to be practical with a scoop made of metal, or even the lightest strong materials like carbon or boron nitride fibers, so would have to involve some kind of electromagnetic scoop, which would restrict the ship to collecting ions, rather than atomic hydrogen. The resistance provided by an immense scoop is also likely to restrict the acceleration of the craft to a lot less than 1 g, limiting the viability of the ramjet design.

If enough hydrogen could be obtained it also isn't clear that we could put together a fusion reactor that would work with the compression produced by a moving ship alone. We struggle to get fusion going with all the resources we can throw at it in a huge reactor on Earth, and reacting hydrogen, the only fuel in sufficient quantity out in space is much harder than the heavier isotopes currently in use. Hydrogen requires star-like temperatures and pressures to make fusion possible. In these circumstances it seems highly unlikely a Bussard ramjet could succeed. While some have suggested modifications including catalyzed nuclear reactions and using lasers to ionize hydrogen atoms ahead of the scoop, this still remains a highly unlikely engineering challenge. It is certainly far more technically difficult than the most dramatic nuclear pulse drive that could be considered.

THE MASS DRIVER BOOST

What could be considered to help a starship on its way is an extension of the mass driver idea (see page 59), which uses a series of electromagnets to get a payload to escape velocity for a relatively weak gravity well like the Moon's. There is no reason why such an electromagnetic linear accelerator needs to be based on a planet, although Newton's third law must be dealt with. In fact, there are some distinct advantages to constructing a mass driver in space, accelerating a spaceship that is not starting from a planet's surface.

The ground-based mass driver is inevitably limited in length by the curvature of the planet—and becomes extremely massive and resource intensive to build as it gets bigger. In space, there is no limit to the length of a mass driver accelerator other than the need to get electrical power to each accelerating electromagnet, and there is very limited need for heavyweight linking structures. It is necessary to have something to tie the different sections together—otherwise the slightest impact could send one drifting out of line—but this could be achieved, even with a virtual structure based on a laser alignment and ion thrusters that countered any drift.

If the space-based mass driver were dispatching a nonfragile cargo (which definitely doesn't include human beings), it could potentially produce a vast acceleration of hundreds of g, getting a probe up to a reasonable fraction of the speed of light while still within the solar system. For a human ship the acceleration would have to be a lot gentler and the electromagnetic "cannon" could only ever be considered as a secondary propulsion device to give an initial boost to a ship that then used, for example, electric ion thrusters (see page 57), but that would not have to carry anywhere near as much fuel as it otherwise would because of that boost. It seems highly likely that any interstellar craft will need to make use of a hybrid launch system of some sort, rather than relying on a single technology.

UNFURLING THE SOLAR SAILS

There is another way to achieve thrust without carrying fuel, one that has a natural appeal to a world that is being constantly told that green energy from solar power is much better for us than the use of oil or gas. That is to make use of the power of the Sun. The Sun pushes out a vast amount of energy, mostly as light, but also in part in streams of particles. Both the light and the particles can be used to push on so-called solar sails, providing a gentle acceleration to a ship. Admittedly this couldn't easily be combined with the electromagnetic cannon—at least the sails could only be unfurled after leaving the mass driver—but it could be seen as an alternative, or to be used at the other end of a journey, where the infrastructure won't be available.

It seems pretty obvious that "solid" particles would be able to provide pressure on sails, like air molecules pushing on traditional sails, but it is less clear why light is able to do this. We don't exactly feel a sudden push every time the Sun comes out. We don't generally notice the effect because light's pressure is very small, but it is present. There is a toy called a Crookes radiometer that is sometimes used to demonstrate light pressure in action. This is rather unfortunate, as the Crookes radiometer doesn't work by the pressure of light at all.

The device (an example can be seen in action at www.universe insideyou.com/experiments.html) looks a bit like an old-fashioned lightbulb, but instead of a glowing filament in the middle it has a set of paddles, arranged around a lightweight axle that runs up the center of the bulb. Each paddle is white on one side and black on the other. When you put the radiometer in a strong source of light, the paddles start to rotate in the partly evacuated bulb. This is supposed to demonstrate light pressure because that light bounces off the white side but is absorbed by the black side. The result should be that the paddles start to rotate black side first. However, in practice, the paddles move in the opposite direction.

The reason this happens is that the real cause of the rotation is that the black sides are warming up more than the white sides, because the black sides are absorbing more photons. The heated paddles warm the air adjacent to the black sides, speeding up the nearby air molecules. Although the bulb is partly evacuated, it does still contain a fair amount of gas. There is more impact from the quicker-moving air molecules, so the paddles start to move, white side first. There is light pressure, but this thermal effect dominates.

When we do see light pressure, it occurs despite the photons of light having no mass. It was predicted by electromagnetic theory before special relativity came along, but also emerges from that most famous of equations $E = mc^2$. The familiar high school formula for momentum, the "oomph" of a moving particle that is felt if it hits you, is mv—its mass times its velocity. If we replace that m with E/c^2 from Einstein's formula we get the value of momentum as Ev/c^2—in the case of light, its velocity, v *is* the same as c (the speed of light). So the momentum comes down to E/c. Each photon of light carries a tiny amount of energy, and that energy produces an even tinier push of momentum just as if it had mass. However, the Sun is pushing out a whole lot of photons, and with big enough sails—we are talking potentially a set that are kilometers across—it is possible to get enough solar pressure to get a ship moving.

Some small-scale experiments have been done to at least demonstrate the principle of solar sails. They need to be constructed out of very thin, lightweight material that is not transparent. These are often based on an extremely thin plastic film with a reflective metallic coating, like aluminum, on the side that will face toward the Sun, though it may also be possible to use stronger but still light materials like carbon fibers to maximize the strength of these flimsy constructions. Sails are often envisaged to be kilometers across to increase the very slight pressure available, though the first-working examples, which were launched in 2010—IKAROS from Japan and NanoSail-D2 from NASA had sails that were just 20 meters (66 feet) and 3.1 meters (10 feet) across, respectively. These sails, though, were merely designed to test the principle, not to propel their ships.

ENGAGE LASERS

This is all fine and dandy, but on its own a solar sail can't give enough acceleration to get a starship up to speed. The answer? It might be to give the Sun a helping hand. After all, we can't produce an actual star, but we can produce light that is a lot more concentrated than sunlight, packing more of a punch in terms of light pressure, using high power lasers. Weighing up the balance sheet, the plus side is very impressive. Remember the Daedalus starship? It could lose 98 percent of its mass if it didn't have to carry its fuel and engine. And as the energy required to get a ship moving is directly proportional to the mass, that is 98 percent less energy required up front. Admittedly, when Daedalus flew it would get lighter as it used up its fuel, but even with an average half the fuel it would still mean a saving of over 50 percent of the energy requirement for the whole trip.

There are some negatives to laser-accelerated sail ships, though. Clearly the technology needs a lot of development. We don't have powerful enough lasers yet. There are some health and safety issues too, with pumping vast quantities of energy out into space that make it likely that the laser would have to be based on the Moon or another off-Earth location. (This would also help as it would remove the energy loss caused by the light scattering from air molecules.) Arguably such a system would have to be based on Mars, as on the Moon or in Earth orbit it could be used as a weapon if directed toward the Earth, making it politically inacceptable. And then there is the sheer duration of the blast. It is likely it would have to be kept up for several years, which implies having very robust technology, being able to sustain a huge drain on the power grid over a long period, and political stability to prevent the plug being pulled partway through. If a new administration comes in and throws out the old ideas, it could be curtains for such a project. There is also the question, as we will see, of what to do when it's time to slow down the laser-boosted craft.

This all seems highly impractical, but as early as 1969, scientists of the paper NASA persuasion were putting forward a theoretical description of how such a flight might be made. The plan would be to site a solar farm in space relatively near the Sun, soaking up large quantities of light. This would be used to power a vast laser. The beam would have to be many kilometers wide to still be useful when a probe has reached a good number of light-years away and the power output would need to be up at the petawatt level (millions of billions of watts)—not much less than the entire Earth's current consumption. Such a beam could get a ship up to a useful percentage of the speed of light in twenty to thirty years.

THERE AND BACK AGAIN

This is all very well (if very long term and dramatic in engineering terms), but how would such a ship turn around and come back? We can't rely on there being an existing beam coming the other way. We can hardly send out engineers ahead of time to build one. The 1969 plan envisaged using the background magnetic field expected to be present at the destination. By deploying a series of charged cables, a turning force would be generated on the ship, as electrical charges moved through a magnetic field produce a sideways force. The ship would very gradually turn in a vast arc, light-years across and could then approach a target system as it headed back toward Earth. Finally the lasers would be deployed again to slow the probe down.

This approach would result in a round-trip time of under 100 years to reach a target in a rough 10 light-years radius, but it could only ever work for an extremely high-speed flypast at thousands of miles per second. The system would have no way of enabling the ship to slow down at its destination, so it would be a huge amount of effort for a very quick "Hello . . . good-bye!" And to make matters worse, this all relies on extreme precision. Managing to fly past, say, Alpha Centauri, or Barnard's star requires vastly more accuracy than a trip around the Moon. Just a tiny error, for instance, in the estimation of the background magnetic field strength when the turning cables are deployed and the probe will neither pass its target nor head back toward Earth. In fact the need to get the ship back into the very narrow (in terms of interstellar distances) laser beam to slow down would make cutting a human hair in half from a few miles away look easy.

Some, including the physicist Freeman Dyson who was involved in the Orion project, have suggested that it would be better to go for a one-way mission, using the interaction with the magnetic field as a brake rather than purely to turn—such a mission would certainly take out some of the vast uncertainty in the original design. It might also be possible to consider some kind of hybrid ship, for instance a version of an electric rocket that uses ions captured by a magnetic scoop, like the Bussard ramjet, but that is powered by a beam from Earth for at least part of its journey, significantly reducing the fuel requirement.

There are, in fact, a whole host of designs and combinations—almost an embarrassment of riches, many of which can be ignored in the short term as being technically impractical. We are not, for instance, going to be building vast banks of lasers near the Sun anytime soon. So it is arguable that there are two real essentials—to choose the best technology that we could use in the next ten to twenty years, and to consider the best options that would probably be practical around a hundred years out.

It is arguable that in the near future enhanced existing technology is the answer. Rockets to get material into space and then, perhaps, a combination of nuclear-powered ion thrusters with a mass driver and/or solar sails. Looking further into the future, some version of nuclear pulse propulsion is still likely to offer the best ability to get up to appropriate interstellar speeds.

ENTER STANDARD ORBIT, MR. SULU

The first thing that obsesses anyone thinking of interstellar travel is how to get there. How the ship is to be powered. But an equally important concern is what to do when a probe does reach its destination. If we are to use the model of what has happened in solar system exploration, we could expect three broad types of approach—a flypast, an orbiting satellite that can send back data potentially over a long period and a landing. All these assume that the robotic probe doesn't come back. In principle any of these expeditions could return—bringing incredibly valuable physical evidence to back up anything that could be beamed by radio. But bringing a probe back could easily double the challenges for the mission.

In fact, even sending radio messages becomes nontrivial over the distances involved. Not only is there the need to be able to beam a message across vast distances, calling for high-power transmitters, there is the time delay. Every light-year in distance means that radio, traveling at the speed of light, will take another year to arrive. Each message from a star, say, 20 light-years distant will be delayed by 20 years on its journey.

The flypast is the simplest approach—and even offers the best possibility of a return of a satellite if it loops around the distant star or planet like a slingshot as in the laser-powered probe (see page 222), and carries on back at the same, or even greater, velocity (assuming the target is traveling in the right direction) without the need to use any fuel. This approach has served us well with our automated exploration to date of the outer planets in the solar system, where slingshot action around the planets has added plenty of speed to the probes, but it can be frustrating as it only gives a narrow window in which to observe the planet. This is bad enough when passing by Jupiter or Saturn, but it would be hugely frustrating if we have gone to all the effort of getting a probe way out to a distant star.

It seems obvious that it would be more attractive to put the probe into orbit, so that the extra-solar planet can be studied for weeks, months, or years, feeding back data as it orbits. Yet this seemingly obvious tactic, all too familiar from the fictional voyages of *Star Trek*'s Starship *Enterprise,* is likely to be painfully costly for a real starship. The problem, which might seem trivial, is stopping, or at least slowing down sufficiently to be able to enter orbit. To achieve any reasonable timeframe on an interstellar voyage, we need to get the ship up to high speed. And slowing down takes just as much energy as getting up to that speed in the first place.

Once the probe has been reduced to the sort of snail's pace we are used to at the moment, then it can use the way it orbits to lose some more speed, as is envisaged for the returning Inspiration Mars capsule—but there will still be the need to remove a huge amount of kinetic energy before that is possible. Solar sails and the like will certainly be useful, but anything that involved carrying fuel to be used at the end of the journey has a huge overhead, as all that fuel adds to the mass of the probe.

Of course, landing on the surface at the destination adds even greater problems. Not only do you have to slow down, you also have to then fight against the planet's gravity as the probe descends. If the planet has air—which hopefully we will be able to determine ahead of departure by the time we are building interplanetary probes—this can be used to slow the craft down, just as we do with a reentry on Earth, but if it is a relatively airless destination like Mars, the only way to land is to deploy rockets or similar technology to control the descent.

To make matters worse, landing brings another significant problem. The onboard computers need to be sufficiently intelligent to select a safe landing site. After all, it would be more than embarrassing to send a probe across light-years for decades, only to have it sink beneath an ocean on its arrival like a cell phone dropped down the toilet. That should be reasonably easy to prevent. But it could be harder, for instance, for a computer to be sure that it is landing on the ground, rather than in some alien tree canopy that would cause the probe to crash destructively through to the forest floor below. And it would be more than a little unfortunate in a number of ways if the planet proved to be inhabited by intelligent life-forms and your onboard computer mistook the local equivalent of a kindergarten for a nice level bit of ground.

WHERE TO GO?

If we have the technology to get to a distant star and manage to slow down on arrival, the next question to address is which star to aim for. While we may well initially target the Alpha Centauri star system, which includes our nearest neighbor Proxima Centauri at around 4.2 light-years, there is little point in this exercise apart from "just getting there" for the glory of heading for another star. What we want to make the journey worthwhile is a star with planets—and better still, with planets that are the right kind of distance from the star and the right kind of size to support life. This was emphasized by a meeting of the U.S. House of Representatives Subcommittees on Space and Research in May 2013, when the chairman, Lamar Smith commented:

> The search for exoplanets and Earth-like planets is a relatively new but inspiring area of space exploration. Scientists are discovering new kinds of solar systems in our own galaxy that we never knew existed. In the universe, is there another place like home? Because of NASA's Kepler mission, we know the likely answer is yes. Imagine how the discovery of life outside our solar system would alter our priorities for space exploration and how we view our place in the universe.

Every year we are adding more and more stars to the catalog of known locations with planets, and we are increasingly able to add data about the size of these planets and how far out from the star they orbit. As soon as it was realized that the Sun was a star there was speculation whether there were planets elsewhere in the universe, and which stars they orbited, leading to the search for stars that seemed to have some kind of wobble that could be caused by planetary motion around it. Early suspicions pointed to the binary star system 70 Ophiuchi, while in the mid-twentieth century Barnard's Star, also in Ophiuchus and one of our nearest neighbors at around 6 light-years, became the favorite—hence it being the proposed target for the hypothetical Daedalus mission (see page 213).

Both these early suspects have more recently been shown to be highly unlikely to host planets, but since the early 1990s a growing catalog of planets has been discovered. At the time of writing, 892 extra-solar planets have been cataloged with more being added to the register every month. Some of the early discoveries were atypical, but the first definitive planet around a conventional main sequence star turned up in 1995, orbiting 61 Pegasi. This was a large Jupiter-sized planet as the early discoveries tended to be. These are usually "hot Jupiters" which are close to the star unlike our own gas giants, and hence more likely to influence its motion and be detectable.

Early methods were indirect, relying on the way the planet's orbit caused the star to wobble. Technically a planet doesn't orbit a star, but rather both the star and its planets orbit their joint center of gravity. Because the star is a lot more massive than even a giant planet like Jupiter, this center is usually contained inside the star, but away from its center, generating a slight wobbly motion. Another indirect approach was to look out for regular dimming in the star's output, when a large planet crossed between the telescope's view and the star, reducing the light output. As our telescopes got ever-more powerful, since the mid-2000s we have also had a few direct observations of extra-solar planets. They are very faint, but if the star itself is blocked out to avoid glare they can sometimes be spotted.

One of the advantages of the direct approach is the spectrum of the light reflected by the planet can be analyzed, giving an indication of its chemical makeup. This can also be possible when a planet is in transit across the star's surface, but is much harder to do under those circumstances. At the moment the vast majority of planets have been detected indirectly, but direct observations are liable to be more practical as new, more powerful telescopes come on line.

Of the planets known at the time of writing, the two nearest stars that seem likely candidates for an expedition are Tau Ceti (coincidentally a favorite destination for many early science fiction novels) and Gliese 581. These are around 12 and 20 light-years away, respectively, so a considerably greater distance than the Alpha Centauri group, but of so much greater interest that it seems likely that they will be considered as likely candidates—meaning that the extra distance and time factor has to be brought into any sensible consideration of where to aim our probes.

ICARUS VERSUS TIN TIN

The most advanced of current proposals for unmanned interstellar probes come from the Icarus Interstellar project. As the name hints, this hypothetical mission returns to the concept of Daedalus (see page 213), but gives the original design a modern twist. As with Daedalus, Icarus is not an actual program to construct a probe, but rather a concept-proving forum to put together the technology, which may one day make interstellar flight a reality, with the aim of achieving this before 2100. Although it is also looking at communications, shielding, and more, the most significant contribution Icarus is likely to make is in the development of fusion engines that could take a probe up to around 10 percent of the speed of light.

Traveling at 0.1 c would make the journey to the nearest stars take in the region of 50 to 80 years—a much more manageable figure than the 75,000 years of *Voyager 1*. At the moment Icarus is primarily a talking shop, more concerned with organizing conferences than actually doing anything. It is very much the independent equivalent of paper NASA. Its communications can, frankly, look like the work of career civil servants. Yet if it manages to push forward work on starship propulsion, it will have achieved a significant part of its goal.

Icarus is a long-term project, depending on the development of sophisticated fusion motors, but in the meantime there is a much closer possibility for an interstellar probe, using existing technology. This represents a parallel development that long term could be the seed of the best possible way to provide large-scale automated exploration of the galaxy. It is called project Tin Tin. There are already a number of satellites in low-Earth orbit known as cubesats or nanosats—very small satellites under 10 kilograms (22 pounds) in mass, which use cheap and dependable off-the-shelf commercial technology, rather than the hugely expensive (and often less reliable) bespoke technology that typifies past NASA launches. Tin Tin builds on this humble technology to commence a program as early as 2015 that would be working toward an interstellar mission, probably initially to the nearest star system, Alpha Centauri.

The aim of Tin Tin is to put the foundations in place from an engineering and technology viewpoint for a series of nanosats. These would work up from local missions, including exploring the Oort Cloud on the outskirts of the solar system, to the eventual aim of dispatching a probe toward the Alpha Centauri system by around 2020. Unlike most interstellar missions, Tin Tin is intended to be a leisurely voyage of around 25,000 years—the only target the project has on speed is to comfortably beat *Voyager 1*'s 75,000-year rate.

Realistically if the Tin Tin probes are launched toward stars they are very likely to be overtaken by later ships. This is a scenario that has been played out a number of times in science fiction where advances in technology mean that a more recent ship can overtake a much-older model. (Think of an airliner taking off the day after an ocean liner crossing the Atlantic for an earthbound equivalent.) For example in Robert Heinlein's *Time for the Stars*, a sublight speed ship, far from home, that has taken many years to get to its location is rescued by a faster-than-light ship that was developed after the travelers departed. This situation is exaggerated by the time dilation aspect of special relativity. While the first ship is in motion, its clocks run slow compared to the Earth, so by the time such a ship returns home, the astronauts have moved into the Earth's future. This gives more opportunity for faster-than-light ships to be built on Earth.

At the extreme, if faster-than-light travel is made possible, practically any conventional starship will arrive at its destination to find faster-than-light travelers are already there. But faster than light is not necessary to beat the Tin Tin probe—just a drive that can take a ship to Alpha Centauri in a humanly comprehensible timescale. That is almost certain to happen. Yet it doesn't make Tin Tin obsolete, because the lessons learned—and the impressive kick up the imagination that is likely to result from seeing a probe leave Earth heading for another star—will more than compensate for Tin Tin's tardy arrival.

It is envisaged that Tin Tin probes would be powered by ion thrusters, in some cases supplemented by solar sails. As we've seen, ion thrusters are already in wide use today, using electromagnetism to accelerate a plasma—a collection of charged particles—which is shot out of the motor to provide thrust. Solar sails have the advantage of adding little mass to the probe and so are ideal for a low-mass nanosat. A variant on the solar sail, known as an electric solar wind sail, could also be used—this is an array of long filaments, like the seed head of a dandelion, which are given an electrical charge and so repel the charged particles in the solar wind, pushing the probe away from the Sun.

Because of the limited size of a nanosat, it is envisaged that a Tin Tin interstellar mission may well have several linked "tins"—one with the thrusters, one with a nuclear power source to generate the electrical power needed for the thrusters, and one for communications. Each of these could be fitted in a 10-kilogram nanosat, perhaps linked by a solar sail, which would double as an antenna to help communicate back to Earth.

OUR PROBES ARE BREEDING

The scale of the task facing any attempt to survey the planets of the galaxy, even in our near neighborhood, is daunting. After the first basic probes have been sent out, if a systematic approach is to be taken then we need to consider the "von Neumann probe." Named after the great Hungarian American mathematician and computer pioneer John von Neumann, these unmanned vessels could be small or large, but they would have two essential capabilities—autonomy and replication.

The first part of this is relatively easy to envisage. Any probes we set off on a voyage of discovery to the stars would have to have enough intelligence to be able to act without guidance from the Earth—inevitable at the range they will operate where the minimum round-trip message time is ten to twenty years, and the chances are that the probe could only carry relatively limited communication capability. While the probe would not, as some have suggested, have to have the entire mental capabilities of a human being, it certainly would have to be better at decision-making than a typical computer of today. However, given advances in computing, it is hard to believe that such capabilities will not be possible within the next hundred years.

The second requirement for a von Neumann probe is replication. The idea is fairly simple. The probe needs to be able to make use of the resources of a planet on which it lands to make copies of itself which can then go on to other planets. So, for instance, we would send a probe to planet 1. Say it made ten copies of itself. These would then travel on to the next nearest destinations—planets 2 through 11. If each of these made ten copies of themselves, the next generation could visit 100 planets—and so on. Of course some, perhaps many, of the probes would be lost, but as long as a few got through and managed to reproduce, the swarm of probes would grow and take in an exponentially growing sphere of planets. This is really the only way in which we could envisage managing a large-scale exploration of exoplanets across many stars.

Saying that the probes need to reproduce is one thing. Making it happen is another. That ability to reproduce, which living things manage so effortlessly, is a lot less trivial when you start incorporating the materials and manufacturing necessary for a space probe, from metals and carbon fibers to power sources and complex circuitry. The probe itself might be quite simple, apart from its brain, but its "reproductive organs" would be horribly complex. Just think of the challenges faced by such a probe landing on Earth. First, it could land on the sea or in the middle of a city. But assuming it was programmed to land somewhere more appropriate, it still would have to find the appropriate ores and other raw materials, refine them, and construct the final structure. And finally it must somehow manage to enable its offspring to blast off the planet's surface out of what could be an intense gravity well without all the usual technology for getting a ship off the surface, before navigating to its next destination. A ridiculously complex challenge.

In the end it is true that this is only an engineering problem, not an impossible physical limitation. But it is such a huge engineering problem that it may never be solvable. Some have argued that the absence of such probes arriving at Earth on a regular basis either means that there is no intelligent life in our galactic neighborhood, or that it isn't possible to develop this technology, because otherwise someone somewhere would have done so by now. The alternative view is that they have been developed, but we have either not noticed them (because they arrived in the time of the dinosaurs, say) or have misinterpreted them as a mythical phenomenon. But I am inclined to think that the practicalities would defeat anyone without technologies we can't even imagine at the moment.

There are those who argue that if such probes were to be feasible they would represent a moral hazard, a danger to the universe that we should not release even if we could. After all, some of the planets they landed on might be inhabited and the local population could be badly affected by the probe's visit. For that matter, they could end up as a physical galactic equivalent of a computer worm, spreading from planet to planet and possibly arriving at the same planet many times and causing significant disruption. However these issues can be relatively easily dealt with by ensuring that the "swarm" does not revisit the same planet, giving the probes bee-like hive intelligence. Modeling the probes on a superorganism like bees or termites where individuals act as if they were part of a larger whole would probably be a necessity if they were to continue to function well without much control from Earth.

The probes would also need to have programming that minimized disruption and avoided the possibility that duplication errors during reproduction would enable them to evolve, perhaps to become a conscious, voracious menace consuming swathes of the galaxy. The great science popularizer Carl Sagan believed that any attempt to design von Neumann probes should be banned, precisely because of this ability to get out of control. However if we ever are to get a handle on the wider universe, such a development seems inevitable should the engineering challenges ever be overcome and it might be better to focus on the controls necessary to keep these (very hypothetical!) devices safe, rather than dismissing the concept entirely.

HARNESSING THE POWER OF A BLACK HOLE

Probably the most exotic proposal for powering interstellar probes using sublight speed drives has a name that would feel entirely comfortable in a science fiction novel—it is the Schwarzschild Kugelblitz starship. To unpack that impressive term, Karl Schwarzschild was a German physicist who, in his spare time in the trenches during the First World War, found that Einstein's general relativity equations predicted that a star with concentrated enough mass would warp space-time so much that light could never escape it. Though he didn't used the name (which was coined by John Wheeler in the 1960s) he had come up with the concept of the black hole. Schwarzschild's name would eventually be applied to the size of a black hole's event horizon, the limit of no return that defines the apparent size of a black hole, known as its Schwarzschild radius.

"Kugelblitz" is the German term for "ball lightning," but it has a very special application in astrophysics. It is a theoretical construct—a black hole that is created not from matter, but from light. As we have seen with the phenomenon of light pressure, in relativity, mass and energy are equivalent. In theory, if enough energy is pumped into a small volume of space, that energy should be sufficient to create a micro-black hole with no conventional matter involved. So a Schwarzschild Kugelblitz starship is one powered by a tiny optical black hole, generated by zapping a small volume of space with concentrated laser-created gamma rays.

It might seem that a black hole is the last thing you want aboard a ship. Our classic Hollywood-inspired image of black holes as voracious consumers of anything nearby might lead us to expect that the power source would eat its way through the ship in seconds. But we are considering a very small black hole here with limited gravitational attraction. A black hole only has the same mass, and the same gravitational pull as the matter in it. If the Sun turned into a black hole tomorrow (something it couldn't do without help), the Earth would continue to orbit it. If the Schwarzschild Kugelblitz power source can be confined in some way, it would not present a risk to the ship. Such confinement would probably depend on being able to create a charged black hole—a theoretical concept—which could then be held in place by electromagnetic containment just as we currently do with plasma and charged antimatter.

So in principle we could have a small, charged black hole confined in the engine room of the ship. How do we turn this remarkable entity into a motor? Another myth about black holes is the not entirely surprising suggestion that they are black, giving off no light. Although nothing emerges from within the event horizon, there is, however, potential for plenty to be going on just outside it. With a black hole of a more conventional size, any charged particles accelerated toward the horizon would give off electromagnetic radiation, but for the small black holes envisaged here, the potential power source is Hawking radiation.

Proposed by Stephen Hawking in the 1970s, Hawking radiation is a quantum effect. Quantum theory tells us that apparently empty space is a seething mass of virtual particles that pop into existence and then disappear again before they could even be detected. When this happens near a black hole's event horizon it is possible for one virtual particle of a pair to enter the black hole, while the other is sent off into space, turning virtual particles into actual ones. The net result is a loss of energy from the black hole. You could look at this as the particle being created from the black hole's gravitational energy, or as an effect of conservation of energy. As the escaping particle has positive energy in its mass, the swallowed particle (as seen from a distance—things would look different in the black hole's gravitational field) effectively has negative energy, reducing the overall energy of a black hole.

The result is that a black hole that isn't consuming ordinary matter and energy from outside will gradually lose energy and evaporate. With a tiny black hole this would happen very quickly. From the viewpoint of our starship, a small black hole would be zapping out particles due to Hawking radiation (which also generate photons from thermal radiation) in all directions. Anything heading out of the back of the ship will generate thrust by Newton's third law, while radiation heading into the ship if suitably reflected would giving the ship a further kinetic energy boost.

There are, of course, a lot of ifs and buts attached to the concept of a Schwarzschild Kugelblitz engine. As yet we have not been able to make small black holes using gamma rays—so this would need taking from theory into practice. Confinement is easier said than done. And because such black holes will naturally evaporate we need to be certain that the holes on board have a long enough life to power the ship to its destination. (As these are assumed to be unmanned probes, there is no need for them to be able to return.) The black holes would have to be produced before the ship departs, because in the end they are only a way of storing massive amounts of energy in a small space. If you could run the lasers that produce the black holes in the confinement of a ship you could also use those lasers to power the ship and cut out a loss of efficiency. The only point of creating the black holes is that they should be able to carry the energy during the flight—but this depends on them not evaporating within microseconds.

UPLOADING THE CREW

In principle, as technology develops there is a compromise to be made between robotic probes and the space ark approach described later in this chapter. This is to have robotic probes that contain human minds—so-called mind uploading. (As has frequently occurred in science fiction, you could also have a real brain embedded in a spaceship, but this raises considerable ethical issues, and suffers from many practical problems in terms of keeping it alive.) The idea of mind uploading is to scan a human brain in such detail that it is possible to re-create its structure in software. With the right hardware (far beyond our current computing capabilities but not inconceivable given the explosive speed of increase in computer power), this device would think like a human being with the same problem-solving abilities, yet without the need to tie up human lives in missions that would last for decades or centuries.

A project at the University of Oxford has suggested that mind uploading is not an infeasible goal. The researchers concluded: "It appears feasible within the foreseeable future to store the full connectivity or even multistate compartment models of all neurons in the brain within the working memory of a large computing system." The researchers foresee bigger problems in terms of having sufficient performance to deal with real-time emulation, but believe that something should be possible by the middle of the twenty-first century. Similarly, the need to actually scan the brain at the level of individual neurons is a significant challenge, but not one that they envisage being a problem beyond the middle of the century.

Were this technology available it would be possible to equip an automated craft with a virtual human crew, capable of the kinds of deduction, decision-making and problem-solving expected of a conventional crew, yet without the *Frankenstein* echoes of a cyborg mesh of a human brain with a machine—and equally without the biological needs to keep such living tissue alive.

The uploaded crew could certainly be physically able to survive the duration of a multi-century voyage with good enough engineering. A more interesting challenge is their mental stability.

Presumably an uploaded mind would suffer exactly the same dangers of boredom and instability as a true human crew. It would be embarrassing, to say the least, if the first starship, manned by uploads, powered itself into the star it was visiting as the unhinged cyber-intelligences decided to commit suicide.

It has been suggested that boredom and isolation would not be an issue because it would be possible to have hundreds or even thousands of uploads on board, forming a kind of virtual community. They could even live in a virtual world, like the scenario in the movie *The Matrix*, but with all the entities artificial. Yet there is always the suspicion that things could go wrong. It is true that in their virtual environment it should be possible for these uploaded entities to lead full and interesting lives within the confines of the ship. Yet a lot of careful design would have to go into ensuring that there couldn't be factions developing, or any of the problems that history has shown can and will arise when a group of human beings are forced to live and work together.

At the extreme, we could envisage a virtual cult arising in which one uploaded individual persuades the rest that they are missionaries from God, or even gods themselves. Given the history of the outcomes of such cults on Earth, it would not suggest a bright future. But perhaps this view is too pessimistic. Maybe the uploaded individuals, leading their full and rich simulated lives, could get on in peace and harmony as they cruise the galaxy. But it is a lot to take on trust.

Robotic probes are fine, but they can never fulfill the drive to *be there* on the frontier, experiencing new worlds. One huge barrier to human space travel just as much with unmanned probes is light speed. In the next chapter we will look at the possibilities for breaking the light speed barrier, but it is entirely possible that planning to do so is little more than wishful thinking. It is very possible that we will never be able to exceed the speed of light. If that is the case, then practicalities could limit us sending out a ship at a fraction of light speed. And that is scary given the distances involved.

Mars enthusiast Robert Zubrin points out that traveling to the nearby stars represents to a civilization that has just developed the ability to fly to Mars the same sort of challenge that getting to Mars would have presented to Christopher Columbus. Not only is the ratio of distances roughly comparable, we are looking at a similar leap forward required in technology. This doesn't mean that we have to have a similar five-hundred-year gap between the two events—our technological abilities are speeding up far quicker than that—but it does put the scale of the challenge into context.

SLEEPING YOUR WAY TO THE STARS

Science fiction often resorts to an apparently simple, but in fact deceptively difficult approach to dealing with a very long journey through space. It suggests we ought to sleep through it. Even on the relatively short trip to Jupiter's moons in *2001: A Space Odyssey*, most of the astronauts on board the *Discovery One* resort to cryogenic hibernation leaving the coolly homicidal Hal 9000 computer with most of the responsibility for handling the ship. The idea (provided you don't have an insane computer in charge) is that unlike an ordinary induced coma where patients age normally, the hibernating astronauts will not age on their journey, however long it may be.

This is all very well in theory, but there is little evidence in practice that such technology is viable. It's true that some people, on their death, opt for cryogenic storage in the hope that a future, high-tech civilization will thaw them out, bring them back to life, and cure them of any disease—but this seems to be pure wish fulfillment. A frozen person is not hibernating, like a bear over the winter.

The bear still breathes and consumes energy. Its brain is still active (if in a very sluggish fashion). The bear is still alive. But once a human is frozen, they are dead.

All brain activity stops. All memories and personality are likely to have gone. This is different from when a patient's heart stops for a few seconds or minutes and they are still revived, as they would have some residual brain function. Even if it were possible to restart the heart of a frozen subject, the individual would have experienced disastrous brain damage. Add in the potential for havoc being caused by crystals forming inside cells—something that is a lot easier to avoid in embryos than a full-sized animal—and you have an unrecoverable situation.

This process becomes even harder over time. It is a very different feat to maintain that capability over a period of hundreds of years compared with successfully putting a human being into cryogenic storage and restoring them the next day. The best technology deteriorates with time. Any spaceship left in charge of corpsicle astronauts would need to be able to maintain its own technology and ensure that the crew could be revived at the appropriate time. While it is not absolutely impossible that some form of hibernation could be used, to hugely slow down the metabolism and perhaps enable a living body to survive several times its normal lifespan, such a process is unlikely to involve death and freezing.

ARKS IN SPACE

If cryogenic storage is ruled out, and assuming that we can't go faster than the speed of light, if we are to take explorers and colonies out to the distant stars, by far the most likely possibility is to take a city into space and have ships where whole generations can live and die. If you simply try to take the mental leap from where we are now to the existence of such "generation ships" it seems too much. But one organization believes it is possible to apply elephant-eating techniques to the problem.

There is an old business and time-management aphorism: How do you eat an elephant? The answer: a bite at a time. The point being, if you try to envisage solving the whole—eating an elephant—it seems an insuperable problem; but if you divide it up—consider the needs a meal at a time—you will eventually get there. The 100 Year Starship project takes that approach. Although it uses a provocation of: "What if a mission designed to take humans beyond our solar system had to be launched in the year 2020?" (more on that in a moment), its elephant goal is to be able to reach another solar system within 100 years. Take the whole goal at once and it appears to be impossible, but break it down into its component requirements and it becomes a more manageable task.

The "2020 What If?" makes use of a powerful creativity technique that relies on our ability to sort things out under pressure. There is no suggestion that anyone is going to build a starship by 2020, but the starting position is what if we had to—if the survival of the human race depended on it? Science fiction often uses the "What if?" scenario as a starting point for its stories, and it is a great way to get a better understanding of what the challenges and blockages are to our achieving a long term goal. By starting with that premise, the 100 Year project team is better able to identify just what has to be overcome in the longer term of a real project.

At this early stage, the 100 Year Starship project is doing little more than identifying those issues. Many of the concerns are about survival. With the best sublight technology (see the next chapter for more on the possibilities of traveling faster than light) we would typically be looking at a 50 to 200 year journey to a nearby star system. To be viable such a starship would need to be able to keep a large enough community alive and well both mentally and physically. Because the duration dwarfs anything yet contemplated, including the journey to Mars, it would be a much bigger problem than anything considered before.

As discussed elsewhere there will be radiation issues. And also there is the need to counter the effects of low gravity, leading to muscles (including the heart) wasting and bones losing calcium and becoming more fragile. Clearly artificial gravity from rotation or constant acceleration would be one key requirement for such a ship. But the bigger picture, the bigger challenge, is constructing an artificial community that could survive without pulling itself apart for decades or even centuries.

DESIGNING THE PERFECT SOCIETY

Human beings don't have a good record for building totally isolated small communities with long-term survival potential. Even relatively long-existing isolated communities like the Amish in Pennsylvania have only managed to keep going for a couple of centuries, and they aren't truly isolated while they still suffer from considerable factional issues. Most closed communities fall apart within years, sometimes in disastrous outbreaks of killings.

It is rather ironic in a mission that is driven by the urge to open up a new frontier that those confined to a truly closed community suffer psychologically specifically because they have no frontier. They can't go out and explore the world—discovering locations, experiences, and people that are different from their tiny world. They can't be pioneers because for them there is no "outside" that is accessible. There is nowhere on Earth where this is as true as it would be on a long duration starship. Nowhere to go, nowhere to hide, nowhere to be yourself and to find things anew.

Worse still, in a double twist of the knife, the kind of person who is likely to want to go on a starship journey is the kind of person who needs a new frontier to explore—exactly the wrong kind of person to be cooped up in an enclosed community for decades and quite possibly, for the first generation, for the rest of their lives. This problem is not insuperable. But it would need careful research and a whole lot of support for the onboard community to allow them to communicate and explore in different ways, to blur the boundary and provide a kind of artificial frontier.

Then there is the matter of fuel. Conventional fuel runs out painfully quickly in chemical rockets. At the very least we would need to be looking at nuclear or antimatter-powered engines, though ideally a starship would want to harvest its fuel as it went along. Every problem faced by an unmanned probe would also apply to an ark in space, with the added complexity of keeping hundreds of individuals alive and psychologically well. It is a remarkably big challenge, and it is no surprise that the real hope is that some kind of faster-than-light drive could be built, removing the need for such a massive endeavor.

How much more simple if it were just a matter of, "Warp factor 3 . . . engage!"

10.

BREAKING THE LIGHT BARRIER

||||||||||||||||||||||||||||||||||||||||||||||||||||||||||||||||||||||||||

Nothing is impossible in this world; it is enough to find out the means through which it can be made.

—*The Ways of Space Navigation* (1929)
German physicist and rocket pioneer Hermann Oberth

On a bad day, a physicist will tell you that light speed is that ultimate barrier and that nothing can ever break it. It is all very well for German space pioneer Hermann Oberth to say, "Nothing is impossible," in the quote above, but physics would firmly suggest that he was wrong. It is impossible, for instance, to break conservation laws, like the law of conservation of energy. It isn't possible to pull energy from nowhere. But it is possible to cheat a bit—to borrow energy from one place and use it elsewhere. That's how we get around the second law of thermodynamics, which says that heat always moves from a hotter place to a colder place. While this is true, we can also cheat by putting energy into the system, something a fridge or an air conditioner demonstrates very practically, shifting heat from the cool interior to the warmer outside.

Einstein's special relativity tells us the light speed barrier is just as unbreakable as the second law of thermodynamics, and in this sense Oberth definitely is wrong. You can't move through space faster than the speed of light, period. It is, indeed, impossible. But without getting around this barrier, we are shackled forever to traveling at speeds that are, on a galactic scale, a crawl. It would be like trying to explore the American West using snails to pull carts instead of horses. But some query whether the barrier is truly an insuperable problem. There is still hope. Because here too there is a way to cheat.

LET'S DO THE SPACE WARP AGAIN

While it's true that nothing can move through space faster than light, it is possible to warp space itself, to produce an apparent motion of any speed you like. The standard big bang theory assumes that the universe has expanded much faster than the speed of light, changing the relative positions of objects within space. So if you can find some way to warp space—or break through it entirely to link one location to another—the restriction falls away. Even a snail can get quickly from one end of a long carpet to the other if you twist that carpet so the two ends meet.

In 2003, General Wesley Clark, campaigning to be the Democratic party's presidential candidate, made a statement that some regarded as little more than fantasy. Clark, defending the need for spending on NASA and on the space program, said:

> I think America needs a dream. America needs a space
> program and I was always a believer in the space program.
> But I have a different vision that goes beyond whether we
> are going to build a space plane or whether we're going to
> try to rehab and build a fourth shuttle or something . . . I
> am a believer in the exploration of space. I would like to
> see mankind get off this planet. I'd like to know what's out
> there beyond the solar system. And I think that we need to
> make a deliberate effort to build public support for
> exploration of a new frontier.

Good, straightforward stuff we have seen from plenty of candidates. But then came the surprise. "I still believe in $E = mc^2$," Clark said in his speech to supporters in New Castle, New Hampshire, "but I can't believe that in all of human history, we'll never ever be able to go beyond the speed of light to reach where we want to go. I happen to believe that mankind can do it." It would be interesting indeed what would happen if there were a president who had the same attitude to interstellar travel that Kennedy had toward a Moon landing. Whether or not you believe it is possible, at least having someone with this kind of vision in charge could result in a real shake-up of the space program.

LEARNING FROM UFOS

There are plenty of examples of conspiracy theories that suggest that the U.S. government has already got drives that are capable of interstellar flight—often supposed to have been captured from aliens. In other words, taken from crashed UFOs. Other lovers of conspiracy suggest that there is leading-edge engineering taking place in black operations for the military developing technology that has never been revealed to the general public. A video interview with the alleged former Lockheed Martin engineer Boyd Bushman includes a quote from ex-Lockheed Martin, skunkworks chief Ben Rich: "We already have the means to travel among the stars, but these technologies are locked up in black projects and it would take an act of God to ever get them out to benefit humanity. . . ."

Bushman claims that magnets can be used to counter gravity to some extent, though his experiments demonstrating this on video are amateurish, to say the least, dropping two objects by hand and relying on eyewitness accounts rather than photography to suggest that one falls faster than the other—evidence that even

Galileo considered insufficient, let alone modern science. Bushman also worryingly states that "Gravity has to have a magnetic complement," and seems to think that magnetism and electromagnetism are different fundamental forces, which doesn't say a whole lot for his expertise in the field of physics. Nor is his credibility helped by his apparent enthusiasm to take the infamous Roswell incident at face value.

The suggestion given in such videos and articles is that Bushman and his colleagues worked on antigravity engines, but unfortunately his rambling description of the theory lacks any conviction. Leaving aside demonstrations of apparent antigravity that are nothing more than well-understood electromagnetic effects, the fact remains that such conspiracy theories fall down when faced with logic. As I pointed out in my book *Gravity*, the best reason to suspect that it is highly unlikely that antigravity technology has been kept secret for fifty years is the way that a known super-secret technology, stealth technology, has come into the public domain. The U.S. military spent billions of dollars on the stealth technology that went into aircraft like the B2 and the F117-A. These developments were real examples of what is sometimes referred to as "black developments," off-balance-sheet expenditures that weren't reported in open government accounts.

Yet the fact remains we all now know about stealth technology, which has no great commercial value outside of the military. Antigravity—which we don't see deployed in battlefields around the world—is on a whole different scale of commercial importance. It would make vast amounts of money if it were made commercial. Such phenomenally large amounts of money that it is impossible that it would not have leaked out or been rediscovered independently. A technology like this simply could not be kept locked away behind the hype and military intelligence smokescreens. Even if we had antigravity, there is no reason for assuming that it would make it possible to exceed light speed—but as there is no evidence of antigravity either, that imagined flying saucer technology is not going to help us get to the stars.

TANGLED IN THE TIME STREAM

The concepts of faster-than-light (FTL) travel bring with them paradox and confusion. It is easy to misunderstand the relationship between FTL and time travel. While it is true that if you can travel faster than light there will always be a way that you can move backward in time; it doesn't follow that as soon as you travel faster than light you will automatically start to move backward through the timeframe. In Adam Roberts otherwise beautifully written novel *Jack Glass*, he gets this aspect of the implications of FTL travel drastically wrong, and it is worth analyzing what happens in that example to see what the true implications of real FTL travel will be.

In the book, the protagonist Jack Glass, uses a "faster than light gun." The gun somehow sends a projectile faster than light. The bullet travels through an unfortunate man standing close-by (who is effectively vaporized), through the wall of a bubble environment, through a docked spaceship, and off into space. But the observers see the events occur in the opposite order. First a distant flash (as the projectile drops back below speed of light), then a damaged spaceship, then a breached bubble, and finally the vaporized individual, making it fiendishly difficult to work out what has just happened.

We can see that this just isn't realistic by thinking of a very simple FTL projectile. Imagine we can send a projectile across the room you are in at the moment at any speed we like. (If you are reading this outside, imagine you are in your lounge or your office). We start by sending the projectile very slowly, then in experiment after experiment send it at a faster speed until in the final run it travels from one side of the room to the other instantaneously—far faster than the speed of light. Let's think about what happens to the timing. Even if the projectile took no time at all to get across the room, we don't see things happening in a reverse order. It is on one side of the room (i.e., by the window); then it is on the other (i.e., by the door). If it was a bullet and passed through people along the way, there would be no time reversal.

To produce time travel, by traveling faster than light, we have to combine a way of beating the light speed barrier—a wormhole, say—with a location that has a time differential. So, for example, we might produce a destination where time has run slowly; simple enough to do because special relativity means this happens if you travel at high speeds. We then use the FTL link to catch up with the slow time location, effectively moving back in time. (The reality would be a little more complex because of the symmetry of relativity, but that's the basic idea.) It isn't sufficient just to move FTL to travel back through time.

ALICE THROUGH THE WORMHOLE

Wormholes, are, of course, a science fiction favorite for FTL travel. The concept came to the fore with the novel (and later movie) *Contact* by science-popularizer extraordinaire, Carl Sagan. In the story, an alien intelligence informs humanity how to use an artificial wormhole to travel to distant stars. Sagan got advice from physicist Kip Thorne on how to come up with a realistic way to travel across interstellar distances. A wormhole is a hypothetical warp in reality that is so extreme that it joins two points that are separated in normal space-time. Sometimes known as an Einstein-Rosen bridge, it involves folding a portion of space so it was U-shaped, then sending two funnels from opposite sides of the top of the U toward each other until they meet. By passing through the funnels you jump the entire length of the U.

To set up a wormhole is anything but trivial. It requires a mechanism to break through space-time, which we don't have. So short of somehow employing a black hole, and then applying a kind of negative energy to keep the wormhole open, the chances of creating a man-made wormhole are slim. In principle, the mysterious energy driving the expansion of the universe could do this— but no one knows what it is or how it could be harnessed. It has also been suggested that a very weak form of negative energy is responsible for the Casimir force that pulls two metal plates together when they are very close to each other. This could have the same result, but making the Casimir effect big enough to be useful, then putting it into practice again seems more fantasy than physics. All in all, a wormhole seems both impractical in terms of setting it up between us and a distant star and being able to successfully pass through it.

THE TRUE WARP DRIVE

Wormholes are, frankly, a messy and unsatisfactory way to travel, even if they could be made manageable. However, in 1994, physicist Miguel Alcubierre working at the University of Wales, Cardiff, wrote a paper that described a true warp drive in the form beloved of science fiction and *Star Trek* in particular. And this was a mechanism that didn't require the construction of something as unstable as a wormhole. The device would contract space-time in front of the ship and expand it behind, pushing the ship forward at speeds that are potentially far faster than that of light. This is possible because, as we have seen, relativity does not apply to the expansion and contraction of space and time itself. In effect, the ship would not move at all, it would change the nature of space-time around it.

If such a warp drive were running, it would effectively cause space-time to ripple around the craft, causing the bubble containing the craft to constantly be repositioned despite not strictly moving with respect to the local space. This is not dissimilar to the way that a surfboard is powered through the water by waves, even though the sea itself does not actual move toward the beach. Perhaps the easiest way to visualize how the drive would work is to imagine an ant sitting on a long tape measure made out of an elastic material. Imagine the ant is halfway between points A and B marked on the tape.

We operate the ant warp drive by squishing up the tape ahead of the ant and simultaneously stretching the tape behind the ant. The ant itself does not move, but it now is closer to point B and further from point A. Unlike the tape measure, with space we can repeat this process indefinitely, carrying the ant as far and as fast as we like, provided we have enough energy. The design of Alcubierre's drive implied that the process would not result in time dilation (because technically the spaceship is not moving), nor would it suffer from serious tidal forces as long as the warp bubble was not too near the craft (echoes of the Starship *Enterprise*'s warp nacelles being kept well away from the ship).

Although Alcubierre's paper was highly theoretical, it was taken seriously by NASA physicist Harold "Sonny" White, who in 2012 announced that it might be possible to at least overcome some of the barriers to making such a warp drive work. (Even though often referred to as "Alcubierre drives," it is hard to imagine that the more generic and easier to spell "warp drive" wouldn't be the name of choice.) If that were done, White has suggested that a practical warp drive would make it possible to reach the nearest stars in 2 weeks, while the kind of 6-month journey envisaged for reaching Mars would open up a horizon of around 50 light-years, taking in a whole raft of stars already identified as having planets.

One of the reasons that Alcubierre's original paper passed by with little real excitement is that practically speaking its demands for energy were beyond anything conceivable. The most compact way to store energy is to use a matter/antimatter reaction where the energy contained in the matter is converted into energy according to the famous Einstein equation, $E = mc^2$. It was originally thought that an Alcubierre drive would need the equivalent of Jupiter's mass (1.9×10^{24} tonnes) to be converted into energy to power the ship—hardly a practical possibility. But White believes there is a way around this. By playing around with the relativistic field equations, White devised an alternative form for a warp drive that had a different shape of negative energy ring, and that oscillated the warp bubble. The result was to reduce the amount of energy required to the equivalent of around 725 kilograms (1,600 pounds), a vast reduction.

TRYING OUT THE WARP

All too often this kind of development has been a matter of pure theory, but White is now working on a proof of concept test device, which would use lasers to generate a perturbation in spacetime, measured by an interferometer. Because they are looking for a tiny shift in the route of a photon, indicating warped space, they are making use of a lab at the Johnson Space Center in Houston, Texas, that floats on pneumatic piers to minimize interference from ground vibrations. It's a long way from a real warp drive, though, even if it works.

The biggest hole in the practicality of the device, present from the very start, was the need to produce negative energy, implying either finding some way to harness the Casimir force on a large scale, or to discover exotic matter—a hypothetical substance with negative mass that would have the same effect. While the problem is a lot less overwhelming than the requirements if wormholes were to be used, there is still a need either to make use of an existing phenomenon on a scale that can't currently be envisaged, or to come up with a substance that may well not exist.

Another issue has been identified by Alcubierre, who never saw his idea as anything more than a thought experiment. He has pointed out that it may not be possible to set up the warp bubble, because somehow a signal would have to pass from the ship to the front of the bubble to control the drive—yet with FTL speeds, there is no way a signal could achieve this. White thinks that this may not be an actual issue, but rather a misunderstanding of the nature of the warp bubble, which after all does not move in the conventional sense. Time will tell.

Warp drives and wormholes make great games for physicists to play. They are thought experiments, taking the current theories of the day and stretching them to near breaking point to see what will happen. White is a rarity in thinking that this approach could ever be made practical. But it is foolish to say "never" at the extremes of physics. We have to bear in mind that leading scientists were saying that nuclear power would never be used to generate energy just a handful of years before Enrico Fermi built the first atomic pile and a few years more before the first atomic bomb.

In 1933, Ernest Rutherford, one of the greatest physicists who ever lived and discoverer of the atomic nucleus, commented: "The energy produced by the breaking down of the atom is a very poor kind of thing. Anyone who expects a source of power from the transformation of these atoms is talking moonshine." The current warp drive is purely conceptual, but to dismiss it entirely now is to fall into the same trap as Rutherford did. It is entirely possible that some way will be discovered either to make it a reality or to make use of a totally new physical discovery in the future.

If we ever do achieve FTL travel we also need to consider the implications for the structure and society of our occupied section of the galaxy. Would we see a *Star Trek*–style federation, or the manipulative tentacles of a repressive galactic empire?

FOUNDATION AND EMPIRE

Many classic works of science fiction, from Isaac Asimov's *Foundation* to the Star Wars movies have described the ultimate conquest of space resulting in the founding of a galactic federation or empire. This is impossible without a FTL drive because of the distances involved, but it is interesting to speculate what would happen if we could travel from star to star at the kind of speed that we have traditionally been able to move from continent to continent. There is something fascinating about the idea of a civilization spanning many worlds—playing with the ways that it would organize and considering how it would communicate.

Unless you are prepared to forget the unpleasantness of past human expansion on Earth, where a (relatively) high-technology society moving in on a low-technology indigenous race has generally led to an unhappy outcome for the locals (think of the fate of Native Americans, for instance), then a major factor that may limit the expansion of a galactic empire is whether or not those juicily attractive Earth-like planets we think we may be able to reach with the appropriate starship technology are already inhabited by intelligent beings.

Science fiction may seem like a modern form of storytelling, but for most of history human beings have speculated about what was out there in distant worlds, and whether there were other forms of life. We have already seen how the Greek writer Plutarch described Alexander the Great as distressed at the limits of his own empire because it did not extend to many worlds. Over a century earlier the Roman poet and philosopher Lucretius, in his epic scientific poem *De Rerum Natura*, commented, "it must be admitted there are other worlds in other parts of the universe and other races of men and of wild beasts."

Not surprisingly, most of the early speculative fiction and

would-be factual speculation put intelligent beings on the Moon or on the most obvious planets of Mars and Venus. It was only with relatively modern telescopes and an increasing understanding of the environment on other bodies that it was realized that it was unlikely that there could be intelligent life elsewhere in the solar system. H. G. Wells, that great father of science fiction knew enough about the barren nature of the lunar surface to put his "selenites" under the surface of the Moon—hence the book's often misquoted title *The First Men in the Moon*. But he still thought it was feasible to base his monstrous invaders for *The War of the Worlds* on Mars.

Wells realized that Mars was a hostile environment, speaking of "air much more attenuated than ours," but he did not realize just how harsh the environment was, thinking that it has "air and water, and all that is necessary for the support of animated existence" and that though its oceans have shrunk, they still cover a third of its surface. For Wells, Mars was the dying home of a race that looked toward the lush, well-provided Earth with jealous eyes. In reality, though, there is no evidence as yet even of bacterial life on Mars, and certainly nothing to suggest there was ever intelligent life there.

Venus, the nearest the Earth has to a twin in the solar system, clung on to its image as a steamy jungle planet longer. So, for instance, in Ray Bradbury's *The Illustrated Man*, a sequence with humans on an admittedly unfriendly Venusian surface is possible. However, as we have seen, the probes that were sent to Venus from the 1960s onward discovered a hell of a planet with a carbon dioxide atmosphere, sulfuric acid rain, and temperatures in the region of 450 degrees Celsius (840 degrees Fahrenheit), which makes any form of life pretty well impossible and again rules out any intelligent neighbors.

Increasingly, then, the only hope for finding intelligence out there is among the stars and it is as we venture beyond the solar system that we need to take serious consideration of just how we will interact with any intelligent species we find—whether we will adopt a *Star Trek*–style prime directive of noninterference. Whether, in fact, we have grown up as a race sufficiently to accept that other intelligent living creatures have the same rights as we do. While whales do not have the same type of intelligence as humans it can be instructive to look at how we treat whales to have some concern—in the way that countries like Japan have found ways to get around treaties banning the exploitation of these animals— but also some hope in the way that most countries have managed to agree to and sustain a ban.

KEEPING THE EMPIRE TOGETHER

Assuming we find planets that are suitable for life but not already the habitat of creatures we would not want to displace, there is the possibility of setting up some form of interplanetary civilization. Many of those writing about galactic empires in fiction have been inspired by and are, to some extent, aping the Roman Empire and its well-documented rise and fall. Galactic empires are often portrayed as grown beyond sensible control, collapsing in places into anarchy, and becoming increasingly cruel to their citizens in the attempt to retain control in the face of rebellion.

Groups of worlds, confederations, and the like were common enough in early science fiction, but it was Asimov with his Foundation series that really set the gold standard. A huge number of fictional galactic empires—notably the Empire in the Star Wars movies—owes a debt of gratitude to Asimov's portrayal, influenced strongly by Edward Gibbon's historical masterpiece, *The History of the Decline and Fall of the Roman Empire*.

The term "empire" was once a proud one, but by Asimov's time it was a tainted concept and any grouping of planets that was to be portrayed as largely beneficial could no longer be sensibly called an empire. It had to be given a veneer of democracy by calling it a federation or league (think, for instance, of *Star Trek*'s United Federation of Planets)—in essence a structure not unlike that of the United States, the spiritual home of most Western science fiction—with the planets replacing the states as entities that had a lot of autonomy, but that pulled together where necessary under a unified federal system.

What is not always thought through in science fiction—or glossed over by making using of physically impossible devices—is the difficulty of communicating across and holding together a large-scale confederation or empire. The Roman Empire managed without any speedy means of communication, but modern empires on Earth benefited hugely from the electric telegraph. On *Star Trek* it is normal to simply use some form of unexplained instantaneous communication, but should the warp drive be the way that a galactic empire is assembled, the fastest way by far to send a message would be for information to be carried on a ship. Just because warp drives can cheat the light speed barrier it doesn't mean that there is any way of transmitting information faster than light.

Such a large-scale empire would have to fall back on the limited communications once available to our pre-radio and pre-telegraphy world, using courier vessels (probably unmanned) to carry physical messages. It would not be impossible to run a collection of planets this way, but it would require a considerable amount of local autonomy and would almost inevitably lead to different parts of the federation or empire splitting off and doing their own thing, as policing such a vast structure without any speedy communication would be beyond practical possibility.

It is fun to imagine a galactic federation or empire, but what we have here is, in reality, the sociologists' equivalent of physicists playing with ideas of wormholes. It makes for a great thought experiment but something that is unlikely to come into being even if warp drives did exist. For the moment, in practical terms we need to set our horizons somewhat closer. Our technology keeps us firmly to our near neighborhood in space, but it does not mean that we have to dismiss the final frontier as a goal.

II.

FRONTIER SPIRIT

||||||||||||||||||||||||||||||||||||||||||||||||||

We shall not cease from exploration
And the end of all our exploring
Will be to arrive where we started
And know the place for the first time.

— "Little Gidding," *Four Quartets* (1943)
T. S. Eliot

The frontier spirit has always driven us to explore, whether it is the simple geographical expansion to cover the surface of the Earth or discovering the extreme limits of our understanding through science. There is no doubt we still have plenty to learn on our own planet. And at the moment space travel is painfully expensive. But there is a danger that the decision of whether or not to explore the solar system and beyond will be decided by a head versus heart debate. This is unnecessary and harmful.

HEAD VERSUS HEART

The debate is typified by the views of two American scientists, the Nobel Prize–winning physicist Steven Weinberg and the astronomer-turned-science-popularizer, Neil deGrasse Tyson. Weinberg, the "head" in the debate, points out the way that a major science project we have already heard about, the Superconducting Super Collider

(SSC) was abandoned, because the funds went instead to the ISS. The SSC would have been significantly more powerful than the Large Hadron Collider at CERN, and would have achieved results a good ten years earlier, argues Weinberg. This would have been a major step in fundamental science research.

By comparison, the ISS has now cost the U.S. government ten times as much as the SSC would have done and has yielded nothing of scientific value. All the useful space science, Weinberg points out, has been done using unmanned satellites. "In the days of the cold war," Weinberg commented, "perhaps it really was important to America to be the first country to put a man on the Moon and not let it be Russia, but today I think that really is irrelevant. The United States is not now in competition with any country resembling the Soviet Union and we do not need to show we are technically just as competent as they are. Any argument of national prestige that could have been valid in the 1960s is certainly not valid fifty years later."

By contrast Tyson, the "heart" in the debate, argues passionately for the manned exploration of space. Tyson points out that manned missions into space are essential to raise interest in the public—and without that interest, the funding fails to follow. When dollars are hard to come by the public can easily say, "Why should we be wasting our tax on obscure scientific research?" In essence, Tyson argues for a human presence in space as a massive PR exercise. And he also seems to agree with his interviewer Stephen Colbert who, as we saw on page 89, stressed the need to be able to feel patriotic support for our astronauts. In essence, Tyson is underlining that achievements in the human exploration of space are inevitably a matter of national pride.

In reality, both Weinberg and Tyson are right, but each is only seeing a small part of the picture. What the virtual debate between the two comes down to is establishing the priorities of the science budget. However this will inevitably be bad for space exploration. There is no doubt that Weinberg is right in terms of the science spending. There are far more bangs per buck to be had from unmanned space expeditions, or from earthbound science than can be gained from manned missions. It is pretty well always *impossible* to justify the risk and cost of putting humans into space for scientific purposes.

WHAT IS SPACE EXPLORATION FOR?

However, there is something else, something bigger, that comes through in Tyson's passion for manned exploration of space. As we have seen, going into space is not really a scientific endeavor at all. It may be for any or all of political, commercial, sociological—even spiritual—purposes, but it really isn't too much about science. We need to separate our thinking here. If we need to shift dollars from somewhere else into space exploration, it really is related closer to defense spending than science—it is about doing something that is at the heart of keeping our civilization safe. By making it thriving and fresh.

I have personally undergone a total change of view in the course of writing this book. When I began it, I very much subscribed to the Weinberg viewpoint that putting humans in space capsules was an idiotic move, because of that huge wasted budget that could be given over to science. With a few exceptions, like the shuttle mission that fixed the problems with the Hubble Space Telescope's mirror, humans have contributed a negligible amount of the scientific value of space exploration to date. But, much as I love science, I have come to realize that space exploration is not about the science. Scientists inevitably overvalue the scientific component of any activity, but in reality there is more to life—and in the case of manned space exploration, there is more to making life worth living.

Opening up the new frontier, exploring space, is a fundamental requirement for the future if we are not to see humanity gradually settling into an asset-poor senescence, with fewer and fewer resources and no drive or energy. If we want the human race to thrive and grow, then we need to be reaching out. And, if T. S. Eliot is right in the quote at the opening of the chapter, we can only benefit by receiving a better understanding of ourselves and the Earth. In the long term, exploring space means making Stephen Colbert wrong—because these are going to be missions on a scale where we probably do need to be saying, "Go, Earth!," rather than, "Go, U.S.A.!"—but it does not mean that the United States (or Europe, or China . . .) lacks a huge role to play, nor does it prevent space exploration being a goal that can unite a nation and give it a new drive and hope in the triumphs of its astronauts and missions within the framework of an international program.

At the moment, we can indeed see the urge to explore space at a nationalistic level, clearly coming through both in Colbert's remarks, but also, for instance, in the Inspiration Mars venture, so-pointedly labeled "A Mission for America." And in the Inspiration Mars website sell:

> We created our foundation to inspire Americans to take advantage of this unique window of opportunity to push the envelope of human experience, while reaching out to our youth to expand their views of their own futures in space exploration. Revitalizing interest among our students in Science, Technology, Engineering and Mathematics (STEM) education is a vital part of our overall mission.

Such patriotic fervor is good to get things moving, but in the long term we have to recognize that these ventures are likely to result in colonies that see themselves not as American, but as Martian (or whatever the destination happens to be). We have the shining example of the United States of America itself. What started as a British colony became a much greater entity with its independence, and we are likely to see the same with colonies in space, especially as, unlike their earthbound counterparts, thanks to the Outer Space Treaty, the colonists cannot claim territory for their native land.

FROM NATIONAL TO HUMAN PRIDE

Back in the mid-nineteenth century, American politics was driven by a perceived duty, sometimes referred to as Manifest Destiny, to expand occupation of the continent from coast to coast and some will inevitably see the expansion into space as something that calls up a similar patriotic fervor. But the exploration of space is much more likely to succeed if we can embrace the reality of a *human* destiny, rather than purely one of a particular nation. The space frontier is one for pride in humanity, not in narrow nationalism.

Generally speaking, big science is already largely an international venture, a move that is unlikely to be reversed. The ISS and CERN with its Large Hadron Collider are two good examples of the sort of international cooperation that is likely to be involved in large-scale space missions, whether to form colonies or to venture further out to the stars. This doesn't mean we stop being proud of who we are as a nation—just that we increasingly might be proud of our achievements as humans, as inhabitants of the Earth, not just as a single country. In that sense, the Inspiration Mars's call specifically and solely to Americans shows it to be a sideshow, rather than the main event.

It is easy to mock the famous words of Gene Roddenberry, which were, after all, simply part of the title sequence of a low-budget science fiction TV show—and yet he got so much right. Anyone of my generation who has any soul can't fail to feel their heart race a little when they hear those famous words spoken aloud. It only takes a slight modification of that original, stirring speech to give us the truth of where we need to be.

> *Space: The final frontier. These are the voyages of the people of Earth. Our ongoing mission to explore strange new worlds, to seek out new life and new civilizations, to boldly go where no one has gone before.*

NOTES

I.

NEW PIONEERS

||||||||||||||||||||||||||||||||||||||||||||||

PAGE 3 – The comments by Steven Weinberg on the opportunities remaining after removing "here be dragons" from the maps are from an interview with the author published in the London *Observer*, March 3, 2013.

PAGE 5 – The NASA cost of $10,000 per pound for payload into space is from http://www.nasa.gov/centers/marshall/news/back ground/facts/astp.html_prt.htm.

PAGE 7 – John Grunsfeld's remarks about the survival difficulties of a single-planet species are quoted in Victoria Jaggard, "How to Build a Mars Colony That lasts—Forever," *New Scientist*, May 15, 2013.

PAGE 7 – Stephen Hawking's comments on the need for humanity to spread out into space to survive were quoted in Melissa Breyer,

"Stephen Hawking Predicts the Imminent End of Humanity on Earth," Mother Nature Network, April 11, 2013.

PAGE 9 – Plutarch's observation of Alexander weeping when he considered the infinite unconquered worlds is from Plutarch, *Moralia—On Tranquility of Mind,* trans. Frank Cole Babbitt (Cambridge, MA: Loeb Classical Library, 1939), p. 179.

2.

SPACE OPERA

PAGE 13 – The quote from Jules Verne on the method of surviving the acceleration of a space cannon is from Jules Verne, *From the Earth to the Moon and a Trip Around It* (Fairford, UK: Echo Library, 2010), p. 79.

PAGE 13 – The quote from Jules Verne on strong springs to deaden the shock of departure is from Jules Verne, *From the Earth to the Moon and A Trip Around It* (Fairford, UK: Echo Library, 2010), p. 100.

PAGE 14 – The quote from H. G. Wells on the lack of substances that are not transparent to gravity is from H. G. Wells, *The First Men in the Moon* (Rockville, MD: Arc Manor, 2008), p. 16.

PAGE 22 – The thoughtful futurology from 1929 is from JD Bernal, *The World, the Flesh and the Devil* (New York: Prism Key Press, 2010).

PAGE 25 – The four books in James Blish's *Cities in Flight* sequence are *They Shall Have Stars* (London: Arrow, 1974), *A Life for the Stars* (London: Arrow, 1974), *Earthman, Come Home* (London: Arrow, 1974), and *A Clash of Cymbals* (London: Arrow, 1974).

3.

SEEING FURTHER

PAGE 30 – Information on the Planck satellite from the ESA website at http://www.esa.int/Our_Activities/Space_Science/Planck.

PAGE 30 – Details of the big bang theory and other aspects of cosmology from Brian Clegg, *Before the Big Bang* (New York: St. Martin's Press, 2009).

PAGE 33 – Information on *Voyager 1* from the NASA website: http://www.nasa.gov/vision/universe/solarsystem/voyager_agu.html.

PAGE 35 – Information on Archimedes' *The Sand Reckoner* from T. L. Heath, ed., *The Works of Archimedes* (New York: Dover, 2002), pp. 221–32.

4.

ESCAPING THE WELL

|||||||||||||||||||||||||||||||||||||||||||||||||||||||||||||||||||

PAGE 47 – Information on Roger Bacon and early references to gunpowder in Europe are from Brian Clegg, *The First Scientist* (London: Constable & Robinson, 2003), pp. 46–47.

PAGE 49 – Information on Konstantin Tsiolkovsky from Chris Gainor, *To a Distant Day* (Lincoln: University of Nebraska Press, 2013), pp. 22–23.

PAGE 50 – Details of Robert Goddard's work from Chris Gainor, *To a Distant Day* (Lincoln: University of Nebraska Press, 2013), pp. 36–52.

PAGE 51 – Information on Hermann Oberth from Chris Gainor, *To a Distant Day* (Lincoln: University of Nebraska Press, 2013), pp. 56–69.

PAGE 56 – Details of the Skylon spaceplane and the Reaction Engines SABRE engine from the Reaction Engines website, http://www.reactionengines.co.uk.

PAGE 61 – Information on the space elevator challenge from http://www.spaceward.org/elevator.

PAGE 64 – Doubts about ever building an elevator are expressed at http://io9.com/5984371/why-well-probably-never-build-a-space-elevator.

PAGE 68 – The observation that an interplanetary ship does not need to be aerodynamic or strong comes from Arthur C. Clarke, *The Exploration of Space* (New York: Harper Brothers, 1951), pp. 58–59.

5.

BACKYARD EXPLORERS

||||||||||||||||||||||||||||||||||||||||||||||||||||||||||||||||

PAGE 71 – Information on Korolev and the R-7 rocket from Robert Zimmerman, *Leaving Earth* (Washington, D. C.: Joseph Henry Press, 2003), pp. 2–3.

PAGE 74 – Alan Shepard's assertion that they wanted to send a dog is reported in Ben Miller, *It's Not Rocket Science* (London: Sphere, 2012), p. 238.

PAGE 74 – The text of Kennedy's Rice University Moon speech is given in full at http://er.jsc.nasa.gov/seh/ricetalk.htm

PAGE 76 – The reason for the unintentional rumpled look of the U.S. flag on the Moon is discussed in Cynthia Phillips and Shana Priwer, *Space Exploration for Dummies* (Indianapolis, IN: Wiley Publishing Inc., 2009), p. 135.

PAGE 82 – Stephen Hawking's warning about the danger of making ourselves visible to aliens was made in a series for the Discovery Channel, quoted on the BBC News website, April 25, 2010, http://news.bbc.co.uk/1/hi/8642558.stm.

PAGE 84 – The analysis of a typical failed NASA program is from Giancarlo Genta and Michael Rycroft, *Space, the Final Frontier?* (Cambridge: Cambridge University Press, 2003), pp. 6–7.

PAGE 85 – Freeman Dyson's contrast between "real NASA" and "paper NASA" is described in W. Patrick McCray, *The Visioneers* (Princeton, NJ: Princeton University Press, 2013), p. 61.

PAGE 85 – The assertion that the majority of major missions since the 1970s have been cancelled is from Giancarlo Genta and Michael Rycroft, *Space, the Final Frontier?* (Cambridge: Cambridge University Press, 2003), p. 7.

PAGE 86 – The short story "Allamagoosa" featuring the offog was first published in May 1955 in *Astounding Science Fiction* and is collected in E. F. Russell, *Like Nothing on Earth* (London: Mandarin, 1986).

PAGE 89 – The interview is quoted in Neil deGrasse Tyson, *Space Chronicles: Facing the Ultimate Frontier* (New York: W. W. Norton, 2012) pp. 186–88

PAGE 90 – Daniel Goldin's outburst on hearing of Dennis Tito's space tourism is quoted in the *Guardian*, February 27, 2013

PAGE 90 – Information on the Virgin Galactic commercial spaceflight service at http://www.virgingalactic.com.

PAGE 94 – Information on the *Soyuz 11* disaster in 1971 from Robert Zimmerman, *Leaving Earth* (Washington, D.C.: Joseph Henry Press, 2003), p. 45.

PAGE 95 – The quote from Alex Saltman on the inevitability of an accident is from Sarah Cruddas, "Space Tourism: The Ultimate Boarding Pass," *New Scientist*, October 17, 2012.

PAGE 100 – The near-Earth object collision wiping out the dinosaurs is likened to a Hiroshima blast every second for 120 years in Donald K. Yeomans, *Near-Earth Objects: Finding Them Before They Find Us* (Princeton, NJ: Princeton University Press, 2013), p. 6.

PAGE 101 – Information on the risk of near-Earth object impact and the possible defenses from Donald K. Yeomans, *Near-Earth Objects: Finding Them Before They Find Us* (Princeton, NJ: Princeton University Press, 2013), p. 109–54.

6.

FRONTIER COLONIES

|||||||||||||||||||||||||||||||||||||||||||||||||||||||||||||||||||||||

PAGE 106 – The full text of the Outer Space Treaty of 1967 is available at http://en.wikisource.org/wiki/Outer_Space_Treaty _of_1967.

PAGE 108 – The biographical details and exploratory ideas of Gerard O'Neill are from W. Patrick McCray, *The Visioneers* (Princeton, NJ: Princeton University Press, 2013).

PAGE 112 – *The New York Times* headline about space colonies is from Walter Sullivan, "Proposal for Human Colonies in Space Is Hailed by Scientists as Feasible Now," *New York Times*, May 13, 1974.

PAGE 116 – The limitations on raw materials and resources to be found on the Moon are detailed in Robert Zubrin, *Entering Space* (New York: Tarcher/Putnam, 1999), pp. 80–81.

PAGE 116 – The *Clementine* and *Lunar Prospector* probes and their results are discussed in Robert Zubrin, *Entering Space* (New York: Tarcher/Putnam, 1999), pp. 92–93.

PAGE 118 – Details of the Foster + Partners design for a 3D printed Moon base are from the BBC News website at http://www.bbc.co.uk/news/technology-21293258.

PAGE 123 – The study of Venus in the microwave region is from C. H. Mayer, T. P. McCullough, and R. M. Sloanaker, "Observations of Venus at 3.15-cm Wave Length," *Astrophysical Journal* 127, no. 1 (January 1958): pp. 1–10.

7.

THE RED PLANET

PAGE 126 – Information on background radiation levels on the Earth and the impact of flying from Brian Clegg, *Inflight Science* (London: Icon Books, 2011), pp. 133–35.

PAGE 127 – The suggestion of using feces as a radiation shield is described in Jacob Aron and Lisa Grossman, "Mars Trip to Use Astronaut Poo as Radiation Shield," *New Scientist*, March 1, 2013.

PAGE 128 – The discovery from Chernobyl that fungi can "eat" ionizing radiation is written up in Ekaterina Dadachova et al, "Ionizing Radiation Changes the Electronic Properties of Melanin and Enhances the Growth of Melanized Fungi," *PLOS ONE*, May 2007. doi:10.1371/journal.pone.0000457.

PAGE 130 – Buzz Aldrin's comments on mission to Mars and founding a Martian colony are from an interview with the BBC on June 21, 2013, detailed at http://www.bbc.co.uk/news/business -22974301.

PAGE 131 – Information on Biosphere 2 from Tiffany O'Callaghan, "A Biosphere Reborn," *New Scientist*, July 27, 2013.

PAGE 134 – The possibility of using Kieserite and other magnesium sulfate salts as a source of water on Mars is discussed in the Royal Society of Chemistry Magnesium Sulfate podcast at http://www .rsc.org/chemistryworld/2013/04/magnesium-sulfate-podcast.

PAGE 135 – The figures for the cost of the Pilgrim Fathers' trip in modern terms and the equivalents for a trip to Mars are from Robert Zubrin, *Entering Space* (New York: Tarcher/Putnam, 1999), pp. 21–22.

PAGE 136 – The estimate of $1.75 trillion as the worldwide defense budget in 2012 comes from the Stockholm International Peace Research Institute—see http://www.sipri.org/research/armaments /milex.

PAGE 138 – The Mars Direct program is described in Robert Zubrin, *The Case for Mars* (New York: Touchstone, 1997).

PAGE 143 – The Heinlein novel suggesting the use of an electromagnetic freight catapult as a weapon of mass destruction is Robert Heinlein, *The Moon Is a Harsh Mistress* (London: New English Library, 1971).

PAGE 144 – The two-stage terraforming of Mars is suggested in Robert Zubrin, *Entering Space* (New York: Tarcher/Putnam, 1999), pp. 108–10.

PAGE 146 – Commander Chris Hadfield's comments that we won't go (on manned deep-space missions) tomorrow comes from a BBC interview in February 2012, published on May 13, 2013, at http://www.bbc.co.uk/news/science-environment-22483934.

PAGE 147 – UK Astronomer Royal Martin Rees made his comments on the ISS in a BBC interview published on May 13, 2013, at http://www.bbc.co.uk/news/science-environment-22483934.

PAGE 147 – Information on the response to NASA's asteroid mission from Tariq Malik, "NASA's Asteroid Capture Mission Flooded with Ideas from Private Companies, Non-Profits," Huffington Post, July 28, 2013 at http://www.huffingtonpost.com/2013/07/28/nasa-asteroid-capture-mission-ideas_n_3663966.html

PAGE 149 – Information on Inspiration Mars from the Inspiration Mars website at http://www.inspirationmars.org.

PAGE 151 – Information on the 1999/2000 Lapierre incident during a simulated spaceflight is from NBC News website, James Oberg, "Does Mars Need Women? Russians Say No," at http://www.nbcnews.com/id/6955149/ns/technology_and_science-space/t/does-mars-need-women-russians-say-no.

PAGE 155 – The quote from Bas Lansdorp on the inability to bring humans back from Mars is from a CBC Radio interview with Brent Bambury given in March 2013 and reported at http://www.cbc.ca/news/world/story/2013/03/16/mars-one-live-die-mars.html.

PAGE 156 – Chris Welch's doubts about maintaining public interest reflecting the drop in public approval of Moon missions two years after *Apollo 11* is quoted in Nigel Henbest, "Life on Mars," *New Scientist*, July 13, 2013.

PAGE 157 – Bas Lansdorp's comparison of the landing on Mars with the 1969 Moon landing and admission that he would not travel to Mars on a one-way trip himself is quoted in Nicola Clark, "Reality TV for the Red Planet," *New York Times*, March 8, 2013.

PAGE 159 – Quotes from Gerard 't Hooft and Norbert Kraft on Mars One applicants are from Victoria Jaggard, " 'Big Brother' Applicants Wanted for One-Way Mars Trip," *New Scientist*, April 23, 2013.

8.

THE NEW GOLD RUSH

PAGE 170 – Information on the Planetary Resources unmanned asteroid mining plan is from their website at www.planetaryresources.com.

PAGE 172 – Tim Worstall's analysis of the economics of asteroid mining for rare metals is from "Asteroid Miners Hunt for Platinum, Leave All Common Sense in Glovebox," *Register,* November 24, 2012, at http://www.theregister.co.uk/2012/11/24/planetary_resources/.

PAGE 173 – Eric Anderson of Planetary Resources is quoted on legal aspects of asteroid mining in Paul Marks, "Who Owns Asteroids or the Moon," *New Scientist,* June 4, 2012.

PAGE 174 – Michael Gold is quoted on the weakness of the Outer Space Treaty in Paul Marks, "Who Owns Asteroids or the Moon," *New Scientist,* June 4, 2012.

PAGE 174 – Information on the Deep Space Industries unmanned asteroid mining program is from their website at www.deepspaceindustries.com.

PAGE 175 – David Gump's analysis of the economics of Deep Space Industries' plans is from an e-mail to the author, dated May 29, 2013.

PAGE 177 – Adam Roberts's ideas on the economic value of energy, resources, and people in space-mining environment are from Adam Roberts, *Jack Glass* (London: Orion Books, 2012).

PAGE 180 – The cost estimates for a space-based solar power station are from Robert Zubrin, *Entering Space* (New York: Tarcher/Putnam, 1999), pp. 71–72.

PAGE 185 – The estimated quantities of metals in a typical 1-kilometer asteroid are from John S. Lewis and Ruth A. Lewis, *Space Resources* (New York: Columbia University Press, 1987).

PAGE 186 – The estimate of 5.5 percent increase in risk is based on the ICRP model, which suggests a linear increase in risk with effective dose at around 0.055 percent per rem. This is discussed in the publication *Risks Associated with Ionising Radiations*, ICRP Supporting Guidance 1. Ann. ICRP 22 (1), 1992.

PAGE 189 – Robert Zubrin's estimate of a 50 times greater mass required to get the same payload from Earth as from Mars is in Robert Zubrin, *Entering Space* (New York: Tarcher/Putnam, 1999), p. 149.

PAGE 192 – The James Blish book featuring the difficulty of having vacuum tubes on Jupiter is the 1956 novel *They Shall Have Stars* (London: Arrow, 1974), p. 42.

9.

PROBING THE GALAXY

PAGE 204 – The relative energy densities of a range of fuels are taken from Richard A. Muller, *Physics for Future Presidents* (New York: W. W. Norton, 2008), pp. 65–69.

PAGE 206 – Details of pulsed nuclear propulsion are from G. R. Schmidt, J. A. Bonometti, and P. J. Morton, *Nuclear Pulse Propulsion—Orion and Beyond*, 36th AIAA/ASME/SAE/ASEE Joint Propulsion Conference, July 2000. AIAA 2000–3856.

PAGE 215 – Details of the Bussard ramjet from Eugene Mallove and Gregory Matloff, *The Starflight Handbook* (New York: John Wiley, 1989), pp. 118–23.

PAGE 220 – The early plan for a solar sail-powered interstellar mission, originally by Philip Norem and enhanced by Robert Forward is described in Eugene Mallove and Gregory Matloff, *The Starflight Handbook* (New York: John Wiley, 1989), pp. 72–73.

PAGE 226 – The quote from House Committee Chairman Lamar Smith on extra solar planets is from the Science Committee website at https://science.house.gov/press-release/subcommittees-review-search-earth-planets.

PAGE 228 – Information on Icarus Interstellar from the project website at www.icarusinterstellar.org.

PAGE 229 – Information on Project Tin Tin from Andreas Tziolas, "Project Tin Tin—Interstellar Nano Mission to Alpha Centauri" (63rd International Astronautical Congress, Naples, Italy, 2012).

PAGE 231 – Von Neumann probes are discussed in Giancarlo Genta and Michael Rycroft, *Space, the Final Frontier?* (Cambridge: Cambridge University Press, 2003), pp. 297–99.

PAGE 234 – The Schwarzschild Kugelblitz starship concept is discussed in a presentation by Jeff Lee to Icarus Interstellar's 2013 Starship Congress. The abstract is available at www.icarusinter stellar.org/2013-starship-congress-speaker-announcement-jeff-lee -singularity-propulsion-the-acceleration-curves-of-a-schwarzs child-kugelblitz-starship-2.

PAGE 237 – The Oxford University report on brain uploading is from Anders Sandberg and Nick Bostrom, *Whole Brain Emulation: A Roadmap*, Technical Report (2008), #2008-3, Future of Humanity Institute, Oxford University.

PAGE 239 – Robert Zubrin's suggestion that with today's technology getting to the stars is like Columbus aiming for Mars comes from Robert Zubrin, *Entering Space* (New York: Tarcher/Putnam, 1999). p. 188.

PAGE 241 – Information on the 100 Year Starship project from its website at www.100yss.org.

10.

BREAKING THE LIGHT BARRIER

PAGE 245 – General Wesley Clark's remarks about the space program and FTL travel were widely reported at the time; e.g., Brian McWilliams, "Clark Campaigns at Light Speed," *Wired*, September 30, 2003 at http://www.wired.com/politics/law/news/2003/09 /60629.

PAGE 246 – The video interview with Boyd Bushman can be seen at http://www.youtube.com/watch?v=VzwOFCSFms4.

PAGE 247 – For a full discussion of antigravity technology and why it is extremely unlikely to exist, see Brian Clegg, *Gravity* (New York: St. Martin's Press, 2012), pp. 250–80.

PAGE 248 – Adam Roberts's erroneous ideas on the time implications of FTL travel are in Adam Roberts, *Jack Glass* (London: Orion Books, 2012), pp 354–55.

PAGE 249 – Details of Carl Sagan's interstellar travel wormhole device, inspired by Kip Thorne from Brian Clegg, *How to Build a Time Machine* (New York: St. Martin's Press, 2011), pp 206–10.

PAGE 250 – Miguel Alcubierre's paper first proposing a practical(ish) warp drive was Miguel Alcubierre, "The Warp Drive: Hyper-Fast Travel Within General Relativity," *Classical and Quantum Gravity* 11 (1994): L73–77.

PAGE 251 – Harold White's paper expanding Alcubierre's idea to a more practical warp drive is Harold White, *Warp Field Mechanics 101*, NASA Archive at http://ntrs.nasa.gov/archive/nasa/casi.ntrs .nasa.gov/20110015936_2011016932.pdf.

PAGE 252 – Information on Harold White's experiment from Danny Hakim, "Faster Than the Speed of Light?," *New York Times*, July 22, 2013.

PAGE 253 – Rutherford's dismissal of the possibility of generating atomic power is from the *New York Herald Tribune*, September 12, 1933.

PAGE 254 – The quote from Lucretius on other inhabited worlds is from Titus Lucretius Carus, *De Rerum Natura*, trans. C. H. Sisson (London: Routledge, 2003), p. 72.

PAGE 255 – The environment of Wells's Martian invaders is from H. G. Wells, *The War of the Worlds* (London: Penguin, 1946), p. 10.

II.

FRONTIER SPIRIT

|||||||||||||||||||||||||||||||||||||||||||||||||||||||||

PAGE 259 – Steven Weinberg's comparison of the ISS and the Superconducting Super Collider was made in a telephone interview with the author in February 2013.

PAGE 260 – Neil deGrasse Tyson's views on space exploration are covered in Neil deGrasse Tyson, *Space Chronicles: Facing the Ultimate Frontier* (New York: W. W. Norton, 2012).

INDEX